University of Cambridge Department of Applied Economics

OCCASIONAL PAPER 27

Britain and the Common Market

The Effect of Entry on the Pattern of Manufacturing Production

University of Cambridge Department of Applied Economics
Occasional Papers

1 *White-collar Redundancy: a Case Study*, by Dorothy Wedderburn
2 *Railway workshops; the Problems of Contraction*, by P. Lesley Cook
3 *The Economies of Large-Scale Production in British Industry: an Introductory Study*, by C. Pratten, R.M. Dean and A. Silberston
4 *Redundancy and the Railwaymen*, by Dorothy Wedderburn
5 *Parking Space for Cars: Assessing the Demand*, by G.J. Roth
6 *Wage Trends, Wage Policies and Collective Bargaining: the Problems for Underdeveloped Countries*, by H.A. Turner
7 *Growth, Profitability and Valuation: A Study of United Kingdom Quoted Companies*, by A. Singh and G. Whittington, in collaboration with H.T. Burley
8 *Productivity in Distribution*, by K.D. George
9 *Industrial Relations in a British Car Factory*, by Garfield Clack
10 *Growth Rate Tables*, by H.T. Burley
11 *The Sterling Area, the Commonwealth and World Economic Growth*, by J.O.N. Perkins
12 *Effects of U.K. Direct Investment Overseas: An Interim Report*, by W.B. Reddaway, in collaboration with J.O.N. Perkins, S.J. Potter and C.T. Taylor
13 *Fortran Programs for Economists*, by Lucy Joan Slater
14 *Demarcation Rules in Shipbuilding and Shiprepairing*, by Geoffrey Roberts
15 *Effects of U.K. Direct Investment Overseas: Final Report*, by W.B. Reddaway, in collaboration with S.J. Potter and C.T. Taylor
16 *Productivity and Capital Expenditure in Retailing*, by K.D. George, in collaboration with P.V. Hills
17 *Productivity in Transport*, by B.M. Deakin and T. Seward
18 *Poverty in Britain and the Reform of Social Security*, by A.B. Atkinson
19 *The Return of Gold 1925½ The Formulation of Economic Policy and its Critics*, by D.E. Moggridge
20 *Is Britain Really Strike-Prone?* by H.A. Turner
21 *The Monopolies Commission in Action*, by A. Sutherland
23 *Towards Economic Independence: Papers on the Nationalization of the Copper Industry in Zambia*, by M.L.O. Faber and J.G. Potter
24 *International Big Business* 1957-67, by Robert Rowthorn, in collaboration with S. Hymer
25 *Soviet Planning Today: Proposals for an Optimally Functioning Economic System*, by Michael Ellman
27 *Britain and the Common Market: The Effect of Entry on the Pattern of Manufacturing Production*, by S.S. Han and H.H. Liesner

Britain and the Common Market

The Effect of Entry on the Pattern of Manufacturing Production

by S.S. Han
and H.H. Liesner

CAMBRIDGE
AT THE UNIVERSITY PRESS
1971

Published by
the Syndics of the Cambridge University Press
Bentley House, 200 Euston Road, London NW1 2DB
American Branch: 32 East 57th Street, New York, N.Y.10022

Library of Congress Catalogue Card Number: 73-149430

ISBN: 0 521 08147 5 paperback

Set in cold type by E.W.C. Wilkins & Associates Ltd,
and printed in Great Britain by Alden & Mowbray Ltd,
at the Alden Press, Oxford

Contents

Tables

Trade by Industry: A Detailed Comparison:

Determinants of Trade Diversion by Industry:

Appendix B Supplementary Tables

Charts

Preface

The possibility that Britain may join the European Economic Community (EEC) raises a great many questions for the economist. One of these concerns the consequences of entry for different British industries. Some British producers are likely to be favourably affected, in the sense that output will be higher than it would otherwise have been, whilst for others the outlook would be adverse. Which are the sectors which look promising, and which industries appear likely to find themselves with lower net sales than they would do if the UK remained outside?

The purpose of our research was to attempt to tackle this question. Not surprisingly serious technical and conceptual difficulties were encountered, and the results obtained therefore provide no more than a tentative 'first shot' at the answer. The present paper, which describes the work undertaken and the conclusions derived from it, is thus in the nature of an interim report, and much further work remains to be done, though it would be futile to pretend that even then anything more than a pretty speculative and tentative set of indicators could be obtained.

It may be asked why, in view of our awareness of the limitations of the study, we did not press on to tackle at least some of the issues left outstanding. The reasons were purely personal – 1 October 1970 was a firm deadline for both of us, and the issue before us was therefore either to conclude the research in time for this Occasional Paper to be in virtually its final form by the stated date, or to leave the results obtained unpublished.

It should be made clear at the outset that the study is not designed to contribute to the current debate about the advantages and disadvantages which EEC membership would bestow upon the British economy. Attempts to read welfare implications into particular patterns of specialization are likely to prove unrewarding, given the present state of ignorance of many of the variables concerned.

The layout of the discussion can be explained very quickly. An introductory chapter which sets the scene, raises some macroeconomic issues, and looks at alternative approaches to the problem, is followed, in Chapter 2, by a survey of UK/EEC trade in all manufactures and by a brief examination of the relevant UK/EEC tariffs. Chapter 3 deals at some length with questions of methodology and presentation; in the course of this discussion some general features of the results are reviewed. Chapter 4 sets out the results in detail, and Chapter 5 contains a number of concluding observations and points out ways in which the analysis should be extended.

It seems likely that the subject matter of this study will primarily be of interest to official bodies and to industry. We have therefore made some attempt to write in a way which will make the material comprehensible to the informed layman.

The costs of the research – principally the full-time salary of one of us - was borne by a two-year SSRC grant. We also wish to acknowledge the help received from the typing and computing staff of the DAE, particularly from Mrs. Lilian Silk and Miss Audrey Twyman. A special word of thanks is due to Mr. Eric Bougourd for his careful work in preparing the study for publication. Professor W.B. Reddaway, Director of the DAE during the relevant period, gave us valuable advice on the approach to be adopted and on ways of presentation; above all, he helped us to sort out several of the conceptual problems. We have also benefited from discussions with other colleagues at Cambridge and elsewhere.

S. S. H.

H. H. L.

Cambridge

Autumn, 1970

1 Introduction

Since 1945 Britain's economic relations with the remainder of Western Europe have been an important issue for all British governments. In the early post-war years the chief problems, at least as seen from London, were the removal of quantitative restrictions on intra-European trade and the improvement of payments relationships among Western European countries, and the main vehicle for inter-governmental collaboration was the Organisation for European Economic Cooperation (OEEC). The 1950's saw the establishment, by Belgium, France, Germany, Italy, Luxembourg, and the Netherlands (the 'Six'), of the European Coal and Steel Community (ECSC), the European Economic Community (EEC),[1] and the European Atomic Energy Community (EURATOM), with the ultimate aim of creating a single European economy. The United Kingdom's initial reaction to the invitation to join the new institutions was negative. Instead, the UK in 1957 proposed the creation of a free trade area for all OEEC countries, including the Six. The negotiations failed, and the UK, together with Austria, Denmark, Norway, Portugal, Sweden, and Switzerland, proceeded to set up the European Free Trade Association (EFTA).

However, relations with the Six remained an important policy issue for Britain. Abortive negotiations about the terms on which the country might join the EEC took place from 1961 to 1963, and the possibility of renewing the negotiations was unsuccessfully explored in 1965 – 67. By 1969 prospects for a further attempt by Britain to gain full membership of the EEC – commonly regarded as the cornerstone of any return to non-discriminatory trading conditions in Western Europe generally – had improved considerably, and negotiations between Britain and the Six commenced again in the second half of 1970.

At the time of writing it is not clear whether these negotiations will be successful. The need for detailed study of the economic consequences of British entry is not thereby removed. Britain's accession to the EEC would be followed within a relatively short period of time by the mutual removal of tariffs and of other trade barriers. Moreover, it is likely that in Britain at least various other economic policy instruments, such as taxes and subsidies, would be altered, particularly those affecting the agricultural sector. It is clearly important to analyse the effects of all such measures upon the economies which would constitute the enlarged EEC as well as upon third countries, at least qualitatively and, as far as possible, quantitatively.

The present study is addressed to one specific aspect of the consequences of

1 The abbreviation EEC, which strictly speaking applies to only one of the three Common Market institutions, will hereafter be used as referring to all three Communities, unless the contrary is specifically indicated.

British entry – the problem of specialization in a wider EEC. The abolition of trade barriers between the UK and the EEC may be expected to result in a reallocation of resources in each area away from activities in which producers cannot compete successfully without the shelter of a tariff and towards activities experiencing rising demand from the newly opened markets. Sectors which decline (relatively if not absolutely) would comprise industries which are inefficient in relation to the average level of efficiency in the area concerned – in other words, industries which are at a comparative cost disadvantage with reference to the pattern of costs in the two areas. Expanding sectors, by way of contrast, would contain industries which enjoy positions of comparative cost advantage.[2]

However, UK and EEC industry will be affected not only by the process of trade creation just described, but also by the diversion of trade away from third countries. A given industry in one area – say, the UK – may experience a growth of sales not only, or not so much, because it has a comparative cost advantage and replaces domestic sources of supply in the second area (the EEC), but also because the continued existence of a tariff or other trade barrier against imports from third countries encourages buyers in the EEC to switch away from third country goods and towards imports from the UK.

The question arises how the various productive activities in the UK and the EEC fit into the pattern described in the last two paragraphs. In other words, what is the identity of each area's promising and unpromising economic sectors (*ipso facto* the sectors respectively likely to contract and expand in the other area)? The provision of an answer to this question would clearly be important to industry, in the widest sense of the term, both in Britain and in the EEC, and also to governments. For instance, the British government would wish to have the earliest possible notice of the prospective decline, because of its relatively high costs, of some particular industry, especially if it were predominantly situated in a development region, and, moreover, a labour intensive activity.

The present study can at best make only a small contribution to the solution

2 This account of the impact of free trade upon the distribution of production assumes *inter alia* that comparative costs in the enlarged EEC are by and large not seriously distorted by factors such as private or government monopolies, discriminatory transport policies, taxation, and regional policies. If this assumption is not fulfilled, there will be a case for the adoption of appropriate steps – 'policy harmonization' – by the UK and the EEC. How far and in what fields such policy harmonization may be required is a question of considerable significance as well as of great complexity. For a simplified account of the principles on which such harmonization should be based, as well as of recent experience by the EEC and EFTA, see H.H. Liesner, *Atlantic Harmonisation: Making Free Trade Work*, The Atlantic Trade Study, London, 1968. See also H.G. Johnson, 'The Implications of Free or Freer Trade for the Harmonization of Other Policies', in Johnson, P. Wonnacott, and H. Shibata, *Harmonization of National Economic Policies under Free Trade*, Toronto, 1968, and C.S. Shoup *ed.*, *Fiscal Harmonization in Common Markets*, *New York*, *1967*.

It should perhaps also be added that although the impact of free trade is being discussed in terms of 'industries' or 'sectors' expanding or contracting, the experiences of the firms making up any one industry may well not be identical. Census of production data suggest that within individual industries costs can vary substantially from one firm to the next; moreover, because of market imperfections some firms enjoy a much more sheltered position vis-à-vis imports than do others. At a relatively general level there is, however, no harm in talking about sectors or industries as if they were made up of relatively homogeneous units; the problem of the individual firm is brought into the argument only when this appears important.

of this problem. It does not claim to do more than that because, as discussed in detail below (Chapter 3), the methodology employed in the research, as well as the statistical material used, suffer from a number of weaknesses, with the effect that the results obtained are only very tentative and provisional.

A more fundamental problem is raised by the fact that the study only deals with manufacturing industry – in other words, that the effects of Britain's accession to the EEC upon activities other than manufacturing have not been considered. In principle this limitation not only means that we cannot say anything very definite about important sectors of the economy (which in aggregate are much larger than manufacturing industry), but also that the range of questions that can be answered with respect to manufacturing is seriously impaired.

The source of the difficulty is the fact that in the long run the consequences of British entry for any particular economic activity, like manufacturing, are in part determined by the effects upon other sectors. Assume, for instance, that EEC membership results in a substantial improvement in Britain's balance of trade, vis-à-vis the EEC, in goods other than manufactures and, perhaps particularly important, in services, and that this improvement is not offset by a worsening balance on other accounts. Provided overall balances of payments (i.e., each partner's balance vis-à-vis the world as a whole) had initially been in equilibrium, the UK balance would now be in surplus (and the EEC's balance in deficit), and a rise in British prices relative to EEC prices (and prices in the rest of the world) would be called for to restore a balanced payments position for each area. However, this rise in relative prices would have adverse consequences for the position in the Common Market of all British industries, including those comprising the manufacturing sector.

The opposite conclusion would hold if the effect of EEC entry upon non-manufacturing activities were such that the balance of trade in the items concerned were to worsen appreciably (or if there were other adverse effects upon the UK balance of payments) – in order to restore a balanced payments position relative British prices would have to fall, with beneficial consequences for British manufacturing and other industries.[3]

Similar conclusions are reached if the problem is considered not with reference to the balance of payments, but in terms of the distribution of resources. A substantial rise in the output of non-manufacturing industries (including services) as a result of increased exports or import substitution would imply a relative if not an absolute reduction in the quantity of resources available to the manufacturing sector and thus in the output it could produce; *mutatis mutandis* a decline in non-manufacturing activities would make more resources available to manufacturing industry and thus permit the output of goods for export and/or of import substitutes to be raised.

3 Whether the price adjustment takes the form of British money wage rates being altered relative to those in the EEC, or of a shift in parities, is immaterial to the argument. It should be added, though, that in the absence of an adjustment of relative prices a balance of payments deficit for the EEC or the UK would presumably have to be met by reductions in output and employment, in which case it would no longer be true to say (as is argued in the next paragraph) that the decline of one sector would be matched by the expansion of another.

These considerations point towards a rather pessimistic conclusion regarding the feasibility of research into the effects of EEC entry upon just one sector. In practice the scope for useful work appears to be rather less severely circumscribed. Two main points must be made.

(a) The last few paragraphs have referred to manufacturing as a whole and no account has been taken of the obvious fact that the industry can be subdivided into a large number of components. There is no a priori reason to suppose that the various parts of the manufacturing sector are in a similar competitive position vis-à-vis the EEC, a point which is very much borne out by the research undertaken. In other words, some manufacturing industries are much more likely to be able to expand output and exports than are others. It follows that the neglect of non-manufacturing activities does not prevent an ordering of manufacturing industries with reference to each unit's *relative* position in an enlarged EEC, so that manufacturing industry A, which is higher up in the ranking order than industry B, is more likely to find itself with demand for its goods growing (or less likely to experience falling demand and output) than industry B. As explained in detail in Chapter 3, the main aim of the research has consequently been to draw up such a ranking order.[4]

(b) Whilst it is true that the absence of any comprehensive analysis of the consequences of EEC entry for the British economy prevents any precise assessment of the absolute effects upon any single industry or activity, comparisons between the size of the particular sector we are concerned with – manufacturing – and any 'reasonable' changes in trade balances in non-manufactures or in invisibles suggest that British manufacturing industry as a whole will be neither very favourably nor very unfavourably affected.[5]

Trade in goods other than manufactures is conveniently taken first. In 1968[6] total UK exports and re-exports amounted to roughly £6 100 million; about £5 160 million of this consisted of manufactures, and the remainder (£940 million) of non-manufactures. The corresponding figures on the import side were: total imports £6 910 million, imports of manufactures £3 420 million, and imports of non-manufactures £3 490 million.[7]

The problem is in what way EEC entry will affect the United Kingdom's balance of trade in non-manufactures. With regard to food (roughly 50 per cent of non-manufacturing trade) official estimates indicate a balance of payments effect

4 Chapter 3 also shows that the relationship between the place of a product on the rank order and the direction and extent of the change in demand and output is not as direct as is implied in this paragraph.

5 Similar arguments apply to shifts in the long-term capital account (not specifically referred to below).

6 This is the latest year for which all the statistics required were available at the time of writing. There is no reason to think that the argument in the text would be materially altered if the figures for 1969 – or, for that matter, those for earlier years – were substituted for the 1968 data.

7 Source: *United Kingdom Balance of Payments, 1969*, Table 7. Commodities and transactions not classified according to kind have been grouped with non-manufactures.

ranging from a small improvement (by £35 million) to a large deterioration (by £875 million), depending on factors such as the response of UK agricultural output to higher prices and the United Kingdom's net contribution to the EEC's agricultural support fund (FEOGA).[8] Other non-manufactures can perhaps be ignored, as both the United Kingdom and the EEC admit many of the goods concerned free of duty, so that changes in the trade balances should not be large. Very tentatively one may therefore say that the United Kingdom's balance of trade in non-manufactures will probably worsen as a result of British accession to the EEC, and that the central estimate of the size of this deterioration – a figure in the region of £400 million to £500 million – is certainly large in balance of payments terms.

However, even if it is assumed that the whole of the balance of payments shortfall on non-manufactures is made up by an improvement in the balance of trade on manufactures, this would still not imply more than a relatively minor increase in the output of manufacturing industry. If the argument is based on 1968 – the latest year for which the necessary figures are available – the rise in manufacturing output occasioned by meeting a £400 to £500 million deterioration in the balance of trade in non-manufactures by a corresponding rise in exports of manufactures[9] comes to only about 3 per cent;[10] moreover, this percentage is still biassed in an upward direction, because the full adverse balance of payments effect would not be felt until the mid-1970's at the earliest, by which time manufacturing output would certainly be a good deal larger than it was in 1968.

With regard to invisibles the argument must be even more tentative. In 1968 total credits amounted to about £3 700 million and total debits to £3 280 million. Just under two-thirds of the debits were recorded in transactions with non-sterling countries (including the EEC) but no further geographical breakdown is available. There is thus no real basis for assessing the effects of EEC entry upon the invisible balance. The United Kingdom government's White Paper states that 'there should be a valuable expansion of our invisible earnings as a result

8 See *Britain and the European Communities: An Economic Assessment* (Cmnd 4289), London, H.M.S.O., 1970, paragraphs 35 to 43. As the text implies, the figures quoted include the contribution which Britain might be asked to make to FEOGA; in the most favourable circumstances the *net* payment would amount to only £50 million, whilst the 'worst' figure would be £620 million. The effect on the balance of trade in food excluding the FEOGA payment is stated in the White Paper to fall in the range + £85 million to – £255 million.

9 For the sake of simplicity it is assumed that the improvement in the manufacturing trade balance is wholly due to a rise in exports. There is no reason to believe that the figures would be substantially different if part of the improvement is brought about by increased import substitution in manufacturing products.

10 In order to allow for the import content of the additional output the gross rise in exports required may roughly be put at £600 million. According to the 1963 input-output table (*National Income and Expenditure*, 1968, Table 21) £100 of manufactured exports on average contains £63 of value-added in manufacturing; an additional £600 million of exports would thus mean a rise in value-added in the manufacturing sector of roughly £380 million. This is 2.8 per cent of the provisional estimate of total net output in manufacturing in 1968 – £13,800 million. Source: Board of Trade Journal, 31.12.1969, pp.1758–1769; Orders III and IV have been excluded so as to make the figures reasonably comparable with the definition of manufactured exports.

of membership' (para. 91), but no figure is put forward. However, given the magnitude of the United Kingdom's invisible trade with all non-sterling countries in the item which according to the White Paper seems the most likely to be affected (Other Services – exports £640 million, imports £360 million), it is highly improbable that any improvement in the invisible balance could be so large as to imply any significant change in the output of manufactures, if once again the extreme assumption is made that the consequent adjustment of the balance of payments wholly takes the form of a worse balance in manufactures.

In the light of these considerations, and taking also into account that the projected changes in non-manufactured goods trade and in invisibles should partially offset one another, one may conclude that the effect of EEC entry upon manufacturing industry *as a whole* is not likely to be either very favourable or very unfavourable; neither outcome would be compatible with reasonable equilibrium in the balance of payments. Subject to some qualifications set out in Chapter 3 it then follows that a product group found at the top end of a ranking order of manufactures should not only 'do better' than a product lower down, but should also do well in absolute terms, in the sense that its output will be larger than it would be if the UK did not join the EEC. The opposite would apply to the producers of a commodity placed towards the lower end of the order – EEC entry would affect them not only more unfavourably than it would those whose commodities are higher up in the rank list, but would also mean that the absolute level of output would be below what it would otherwise have been. (Whether this implies that output would actually decline, or simply grow less fast, cannot be deduced from the data.) The absolute position of goods placed away from the two ends of the rank order, and particularly that of goods found towards the centre, must, however, remain indeterminate. The problem is taken up again in Chapter 3.

It appears appropriate to conclude this introduction with a general outline of the methods available for an investigation of the kind attempted in the present study, with some reference to other work. There are at least three distinct ways in which one could set about an investigation of the effects of a major change in trade barriers upon the manufacturing sector – (a) by carrying out a comparative analysis of data relating to the costs and prices of manufactures in the countries concerned; (b) by asking industrialists how they expect the changes in trade barriers to affect their sales both in the domestic market and in the markets of the partner countries; (c) by undertaking a comparative analysis of the foreign trade of the countries concerned in the various commodities produced by manufacturing industry, in the expectation that trade patterns prior to the change in trade barriers are at least broadly indicative of the comparative cost positions of the various manufacturing industries in each country.

The first approach has been followed by the Wonnacotts,[11] whose assessment of the industrial effects of a Canada/United States free trade scheme was based upon a detailed examination of the main factors determining the relative cost positions of Canadian and US manufacturing industries. Moreover, other studies of the industrial consequences for Canada of either a free trade arrangement with

11 R.J. Wonnacott and P. Wonnacott, *Free Trade between the United States and Canada: The Potential Economic Effects*, Harvard University Press, Cambridge (Mass.), 1967.

the United States or of a general 'North Atlantic' free trade area (NAFTA) have relied in the main on comparative cost and price data.[12] Evidence relating to costs and prices and to the general position of the main manufacturing industry groups in each prospective partner country also formed the bulk of the material on which the Economist Intelligence Unit based their 1957 study of the effects on British manufacturing industry of the setting up of a 'Maudling-type' European free trade area comprising the Common Market, the United Kingdom, the Scandinavian countries, Austria and Switzerland.[13]

The industrial survey method appears to have been used much less widely. The Confederation of British Industry's 1966/67 examination of the effect of EEC entry upon British industry[14] relied in part on submissions by certain industrial sectors, but systematic consultation of industrial opinion was only a matter of recommendation to the CBI Council, and the survey actually conducted does not appear to have covered the prospects for particular industries in an enlarged EEC. It might be added that enquiries by the 'little Neddies' into factors determining industrial performance in general and relative competitiveness vis-à-vis other countries in particular have from time to time touched on the effects of EEC membership, but very little by way of solid evidence appears to have been collected.[15]

Foreign trade data were apparently first used as the basis for establishing comparative cost positions in a paper attempting to indicate how different British manufacturing industries would fare in a 'Maudling-type' free trade area.[16] The same approach was adopted by Bela Balassa in an investigation of the industrial consequences of multilateral tariff reductions which might be negotiated among the main industrial countries on a non-discriminatory basis, possibly in the course of GATT's 'Kennedy Round'.[17] A Dutch paper considering

12 W.E. Haviland, N.S. Takacsy, E.M. Cape, *Trade Liberalization and the Canadian Pulp and Paper Industry*, D.E. Bond and R.J. Wonnacott, *Trade Liberalization and the Canadian Furniture Industry*, J. Singer, *Trade Liberalization and the Canadian Steel Industry*, all published for the Private Planning Association of Canada by University of Toronto Press, 1968/69.

13 The Economist Intelligence Unit, *Britain and Europe: A Study of the Effects on British Manufacturing Industry of a Free Trade Area and the Common Market*, London, 1957. The description of the arrangement as 'Maudling-type' reflects the fact that on the British side Mr. Reginald Maudling was in charge of the negotiations.

14 Confederation of British Industry, *Britain and Europe*, vol.1, *An Industrial Appraisal*, vol.2, *Supporting Papers*, vol. 3, *A Programme for Action*, London 1966/67.

15 See, for example, NEDO, *Market – the World: A Study of Success in Exporting*, Report of the Working Party of the Economic Development Committee for the Mechanical Engineering Industry, London, 1968.

16 Cf. H.H. Liesner, 'The European Common Market and British Industry', *Economic Journal*, June 1958, pp.302–316.

17 Cf. B. Balassa, 'Trade Liberalization and "Revealed" Comparative Advantage', *Manchester School*, May 1965, pp.99–123. The results of Balassa's work were also used in a wider project directed by Balassa – see B. Balassa, *Trade Liberalization among Industrial Countries: Objectives and Alternatives*, New York, 1967, and B. Balassa *et al.*, *Studies in Trade Liberalization: Problems and Prospects for the Industrial Countries*, Baltimore, 1967. See also S.J. Wells, *Trade Policies for Britain: A Study in Alternatives*, London, 1966.

the position of Netherlands industry in the EEC also relied on an analysis of foreign trade performance.[18]

Each of the methods outlined has its strong points as well as its disadvantages. The first (the analysis of relative costs and prices) is the most direct and probably the soundest approach, but it runs into serious problems over the data required, because industrial statistics of a high degree of international comparability are needed. In the particular case under examination in the present study these conditions could probably be met only by a census of production-type enquiry covering all the countries involved, a task which would clearly have to be undertaken by an authoritative international body in command of the necessary funds.

The main case in favour of the second method (collecting the views and forecasts of industrialists) is that the people in charge of industry are in possession of a great deal of relevant information not contained in any statistical series, and a properly conducted enquiry should unearth at least some of this evidence. Again, however, many problems arise. As the Industrial Enquiry conducted in connection with Britain's 1965 National Plan indicated, industrialists do not find it easy to make forecasts for conditions with which they are not familiar; moreover, there is the difficulty of obtaining a reasonably consistent set of answers. Detailed questionnaires backed by intensive interviewing could overcome some of these problems (provided the response rate was good), but the cost would be considerable, and indeed prohibitive as far as the present study is concerned.

The use of foreign trade data for the purpose of establishing comparative cost positions has the advantage – a decisive one in the context of this study – that up-to-date and reasonably comparable statistics are readily available and that costs in general are low. At the same time the method also suffers from many imperfections. These are discussed in detail in Chapter 3; for the present it may be concluded that no single approach is likely to be wholly satisfactory and that ideally one would use two or even all three methods in combination.

18 I.E.G. van der Boor and A.F. Veldkamp, 'De concurrentiepositie van 82 Nederlandse uitvoerprodukten bij de aanvang van het in werking treden van de Euromarkt', *De Economist,* April 1959, pp.257–283.

2 United Kingdom and EEC Trade in Manufactures: An Overall View

The main task of the present chapter is to consider the UK's and the EEC's external trade in manufactures as a whole during the period 1965/67 (in the case of the EEC 'external' means trade with non-Community countries, i.e., intra-trade is excluded throughout). The salient facts are summarized in Table 2.1; the upper part of the table is taken first.

To begin with it may be noted that for both prospective partners exports to the other (valued f.o.b.) form a relatively small part of *total* external sales of manufactures and that this is particularly true of the EEC (whose exports to the UK were only just over 7 per cent of total exports, as compared to the 18½ per cent of total British exports accounted for by sales to the EEC). If mutual UK/EEC trade is compared with each partner's total *imports* of manufactures (all trade flows being valued c.i.f.), the resulting proportions (about 28 per cent for the UK and 20 per cent for the EEC) are rather higher – a reflection of the fact that both the UK and the EEC have large positive balances in their manufactures trade.[1]

However, the relevance of the percentages just quoted lies not so much in their relation to each other but rather in their absolute levels. If we take the second set of figures first, it is clear that both the EEC and the UK rely for the bulk of

1 It will be noted that in the case of EEC exports to the UK the f.o.b. and c.i.f. values diverge quite markedly – the c.i.f. figure being about 14 per cent higher than the f.o.b. one – whilst the c.i.f. value for UK exports to the EEC is only 2 per cent above the f.o.b. one. It is normally assumed that the c.i.f. valuation of a trade flow is something like 10 per cent above the f.o.b. figure, and the more surprising set of results is therefore that relating to UK exports to the EEC. It is difficult to know what the reason might be. The under-recording of UK exports in the period 1963–1969 (see *United Kingdom Balance of Payments*, 1969, page 82) cannot be responsible, as the 'correct' data would give an even higher f.o.b.-based figure. A more promising possibility would appear to be errors in the recording of the destination of UK exports. This is indicated by the comparison between UK exports to individual EEC countries (measured f.o.b. and derived from British sources of information) and each EEC country's imports from the UK (measured c.i.f. and based on the recipient's records). The results for France, Germany and Italy are more or less in accordance with what on would expect – the c.i.f.-based figures are significantly above the f.o.b. ones – but in the remaining two cases (Belgium/Luxembourg and the Netherlands) the c.i.f. value is *below* the f.o.b. figure, by $6 million and as much as $48 million respectively. This suggests that some of the goods shipped from the UK to Belgian and especially Dutch ports to be forwarded from there to other destinations – most probably outside Europe – were recorded as UK exports to Belgium/Luxembourg and the Netherlands respectively, but without detailed investigation of the practices of the various parties involved it is not possible to come to a firm conclusion.

It might be added that the detailed examination of trade flows presented in the main part of this study is based on c.i.f. statistics and that the particular potential source of error just referred to would thus not affect any of the results – but this does not, of course, rule out the possibility of other 'bugs' in the data.

Table 2.1 *UK and EEC Trade in Manufactures*

A. Overall Trade

	UK		EEC	
	1965/67 average ($ million)	1965/67 / 1956/58 (per cent)	1965/67 average ($ million)	1965/67 / 1956/58 (per cent)
	(1)	(2)	(3)	(4)
Total exports (f.o.b.)	11,641	155	24,506	209
of which				
to prospective partner	2,150	230	1,761	239
to other European OECD	2,204	218	7,877	245
to third countries	7,287	131	14,868	191
Total imports (c.i.f.)	7,016	274	11,228	258
of which				
from prospective partner	2,004	264	2,193	232
from other European OECD	1,230	324	2,930	245
from third countries	3,782	267	6,104	276

B. Trade by EEC Country of Destination/Origin

	UK exports to EEC (c.i.f.)		EEC exports to UK (f.o.b.)	
Country of destination/origin	1965/67 average ($ million)	% of recipient's total imports	1965/67 average ($ million)	% of country's total exports
	(1)	(2)	(3)	(4)
Belgium/Luxembourg	419	29	226	11
France	484	22	306	6
West Germany	578	14	717	6
Italy	321	21	260	7
Netherlands	391	29	251	14

Notes: (i) Sources: OECD *Commodity Trade*, Series C; UN Statistical Papers, Series D, *Commodity Trade Statistics*; and EEC *Foreign Trade*, Monthly Statistics.

(ii) Trade *among* the EEC countries – 'intra-trade' – is excluded throughout.

their imports on manufactures on non-partner sources of supply. The preferential reduction in trade barriers resulting from British accession would therefore *prima facie* provide plenty of scope for trade diversion – EEC buyers substituting British goods for those so far imported from other countries, and British importers similarly switching demand away from other countries and towards EEC sources of supply. This suggests that detailed analysis of the position of different industries must take at least some account of the effects of additional trade flows being generated through trade diversion. It might be added in passing that although at first sight the possibility of trade diversion should be welcomed, in view of the stimulus given to each prospective partner's exports, it is by no means obvious that in the end either or both partners would derive a net gain, because it is implied that they switch to sources of supply which, before the discriminatory reduction in trade barriers, were regarded as in some sense inferior.[2]

The significance for present purposes of the first set of percentages – which show that for both partners exports to other countries form the bulk of total external sales – is rather less obvious. One possible interpretation would be this – the figures indicate that on average there is plenty of scope for the exporters of both partners, and especially of the EEC, at the margin to switch sales away from third markets and to each other's market if a change in circumstances, such as the abolition of mutual trade barriers, should suggest to them that such action would be worth while. In other words, the hypothesis is that the relatively small size of mutual UK/EEC trade makes it appear improbable that in general capacity limitations would be a serious bar to exporters at least beginning to exploit the new opportunities which will be open to them when trade barriers are removed. Mutual trade may thus be expected to expand quite rapidly once UK accession is achieved.

In this context attention may briefly be drawn to the data in columns 2 and 4, showing the trade levels in 1965/67 as a percentage of those in 1956/58, the last three years of non-discriminatory trading in Western Europe.[3] The figures indicate quite clearly that UK trade with 'other European OECD' – mostly Britain's EFTA partners – had grown much more rapidly than total British trade, or trade with the other areas separately shown, with the one important exception that the rise in exports to EEC was slightly faster than that to 'other OECD'. In the case of the EEC the relevant comparison is that between the Community's trade with the various markets shown in the table and internal trade. In 1965/67 EEC exports of manufactures to member countries were nearly four times the 1956/58 average, a much faster rate of increase than that in the EEC's exports to, or imports from, non-EEC countries. However, it must be pointed out that these results do not really get one very far, because valid conclusions can only be drawn from a comparison between the changes in trade flows as recorded and those which would have taken place in the absence of discriminatory trading conditions. An analysis of that kind is quite beyond the scope of the present study and would in any case be only partially relevant.[4]

2 Analysis of the effects of trade diversion is an important aspect of the theory of customs unions, the subject of a voluminous literature. For summaries of the theory see R.G. Lipsey, 'The Theory of Customs Unions: A General Survey', *Economic Journal*, September 1960, pp.496–513 (reprinted *inter alia* in J. Bhagwati *ed.*, *International Trade*, Penguin Modern Economics Readings, London, 1969), and H. Shibata, 'The Theory of Economic Unions: A Comparative Analysis of Customs Unions, Free Trade Areas, and Tax Unions' in C.S. Shoup *ed.*, *Fiscal Harmonization in Common Markets*, New York, 1967, pp.145–264.

3 The first cut in internal tariffs in the EEC was made on 1st January 1959, with EFTA following suit 18 months later. The establishment in 1953 of the European Coal and Steel Community is disregarded.

4 Work along these lines has been done by Bela Balassa ('Trade Creation and Trade Diversion in the European Common Market', *Economic Journal*, March 1967, pp.1–21), by E.M. Truman,('The European Economic Community: Trade Creation and Trade Diversion', *Yale Economic Essays*, Spring 1969, pp.201–257), and by the EFTA Secretariat (*The Effects of EFTA on the Economies of Member States*, Geneva, 1969). All three studies show that free trade has stimulated trade among the member countries to a significant degree.

In the discussion of Table 2.1 it has so far been implicitly assumed that the United Kingdom would be the only EFTA country to join the EEC. It is probably more realistic to work on the basis that the majority of the remaining EFTA countries will accompany the UK in some form or another. Moreover, at least one other OECD country – Ireland – is likely to follow suit.

A rough way of allowing for this possibility is to compare mutual UK/EEC trade with each partner's total trade *minus* trade with other European OECD countries. On this basis UK imports from the EEC work out at 35 per cent of total British imports, and EEC imports from the UK at 26 per cent of total EEC imports. In the case of exports the figures are somewhat lower – 23 per cent of UK exports (excluding exports to other European OECD) went to the EEC, and 11 per cent of the EEC's exports to the UK. These percentages indicate that scope for trade diversion, and for substitution of markets on the export side, may be somewhat lower than was suggested above, but the argument is clearly not thereby invalidated.

One of the simplifications resorted to in the present study is the treatment of the EEC as one unit. At the aggregate level of trade in manufactures with which we are at present concerned a breakdown of UK exports to the EEC by country of destination within the Community, and of EEC exports to the UK by country of origin, can be handled easily enough, and the figures are shown in the lower part of Table 2.1. Taking UK exports[5] first one finds, not surprisingly, that there is a good deal of variation in the size of UK exports to the five EEC countries (Belgium/Luxembourg being treated as one country for present purposes), with sales to Italy on average less than 60 per cent of those to Germany. However, it seems more useful to express the UK's exports of manufactures to each country as a proportion of that country's total imports of manufactures from non-EEC sources. The resulting percentages (column 2) again differ quite markedly; as much as 29 per cent of Belgian and Dutch imports of manufactures from non-EEC sources are shown as having originated in the UK, as opposed to about 14 per cent of German imports.

These variations can be interpreted in a number of ways. For instance, it may be suggested that 'tastes' in a country like the Netherlands are in some sense more favourable towards UK products than they are, say, in Germany or in Italy, so that UK producers concentrating on the Dutch market are particularly well placed; alternatively, it may be argued that the strong position of British goods in Belgium/Luxembourg and in the Netherlands points to a relatively high degree of exploitation of opportunities there, so that the scope for a further expansion of sales is relatively limited. Either interpretation in effect amounts to saying that the EEC market is by no means homogeneous, which in turn indicates that this study's treatment of the EEC as one unit constitutes a serious weakness. However, lack of resources prevented the adoption of a less aggregated approach.

5 Because of the factors discussed in footnote 1, and also because of the need to make each country's imports from the UK comparable with that country's total imports of manufactures, UK exports are valued c.i.f., i.e., measured as each country's purchases from the UK, rather than f.o.b.

EEC exports to the UK by country of origin similarly show marked variations whether measured in absolute terms or as a proportion of each country's total exports of manufactures to third countries (column 4). In no case, however, do the figures suggest that the earlier argument concerning the ability of producers to switch sales to the UK market may not apply to each EEC country taken separately.

The final point to be noted from Table 2.1 concerns an important methodological question considered in the next chapter. As will then be explained, the detailed analysis of trade performance proceeds both in terms of mutual UK/EEC trade and in terms of each prospective partner's exports to third countries (also referred to as 'neutral markets'), defined as exports to the world *minus* exports to OECD (Europe). It is evident from the figures in the upper part of the table that mutual exports and neutral market exports taken together account for the bulk of total exports of manufactures, particularly in the case of the UK (whose sales to other European OECD countries amounted to less than 20 per cent of her total). This means that an analysis embracing both mutual trade and exports to neutral countries omits only a relatively small part of total export trade in manufactures. (It might be added that the question of coverage concerns not only the markets included, but also the proportion of manufactures trade subjected to detailed scrutiny. This matter is taken up below.)

The second task of this chapter is briefly to consider some features of the UK and EEC tariffs on manufactured goods.[6] Examination of tariffs by industry is dealt with in Chapter 4, but there are a number of general points which are best taken at the present stage, even though some of this discussion anticipates material to be introduced in the next chapter.

The frequency distributions of British and EEC tariffs on the 230 commodities included in the study are shown in Chart 2.1, and the breakdown by SITC section in Table 2.2.[7] Two sets of rates are included in the graph for the Common External Tariff (CET) of the EEC – the broken line refers to the tariffs in force in the period 1965 to 1967, and the 'dot-dash' line to the post-Kennedy Round CET (fully effective as from 1.1.1972). For the time being the post-Kennedy Round CET is disregarded.

It is clear that during the mid-1960's British tariffs even without the addition of the surcharge (see Appendix A) were significantly higher than the CET. The mode of the British tariff lies between 15 and 19.9 per cent, whereas that of the CET lies in the next lower range – 10 to 14.9 per cent. In the case of the UK rates of duty for 76 product groups (33 per cent of the total) are classified as 20 per cent or above; the equivalent figure for the EEC is 23 product groups (10 per cent).

Table 2.2 reveals that the general tendency for British tariffs to exceed EEC rates of duty holds for each of the four SITC sections separately shown.

6 Technical details concerning the derivation of tariff rates are contained in Appendix A.

7 The approach adopted implicitly gives an equal weight to each tariff rate. There are well-known objections to this procedure, but no alternative was available.

Chart 2.1 *Tariffs on Manufactures*

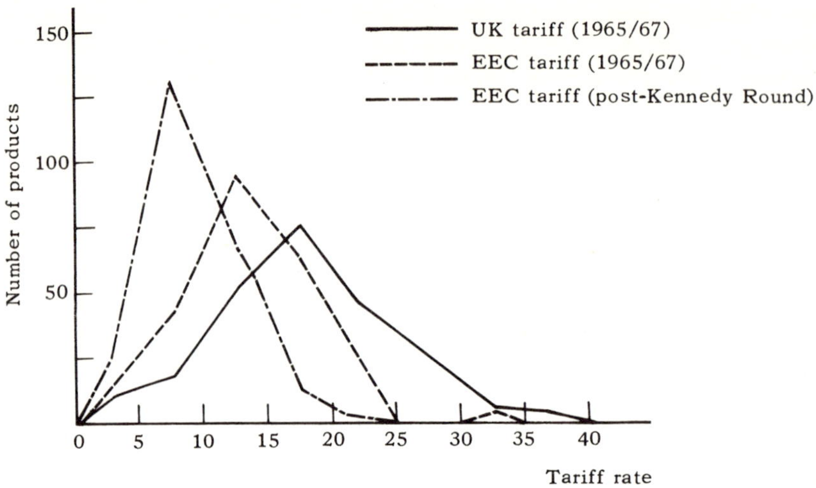

Source: See Appendix A.

Table 2.2 *UK and EEC Tariffs on 230 Manufactures* (1965/67)

Tariff (per cent)	UK 5	UK 6	UK 7	UK 8	UK 5-8	EEC 5	EEC 6	EEC 7	EEC 8	EEC 5-8
0.0-4.9	1	8	0	1	10	2	7	1	1	11
5.0-9.9	6	10	0	2	18	5	30	4	1	40
10.0-14.9	11	22	16	2	51	15	30	38	12	95
15.0-19.9	7	36	24	8	75	14	27	7	14	62
20.0-24.9	2	20	16	6	44	2	7	9	3	21
25.0-29.9	10	5	3	8	26	0	0	0	0	0
30.0-34.9	1	1	0	2	4	0	1	0	0	1
35.0-39.9	0	0	0	2	2	0	0	0	0	0
Total	38	102	59	31	230	38	102	59	31	230

Notes: (i) Source: See Appendix A

(ii) The descriptions of the SITC sections are as follows:
SITC 5 Chemicals
SITC 6 Manufactured goods classified chiefly by material
SITC 7 Machinery and transport equipment
SITC 8 Miscellaneous manufactured articles

Especially noteworthy are perhaps the figures for SITC 5 and 8, because of particularly marked disparities in the number of rates in the higher ranges (20 per cent and above).

The broad similarity of the shapes of the frequency distributions of UK and EEC tariffs in operation in the mid-1960's means that a uniform cut in British tariffs – by 5 per cent – would make the overall picture much the same, because the UK curve would be shifted by one interval to the left. It does not, however, follow that tariff differences on particular products would also be eliminated. As

Chart 2.2 *Tariff Differences*
(UK minus EEC, 1965/67)

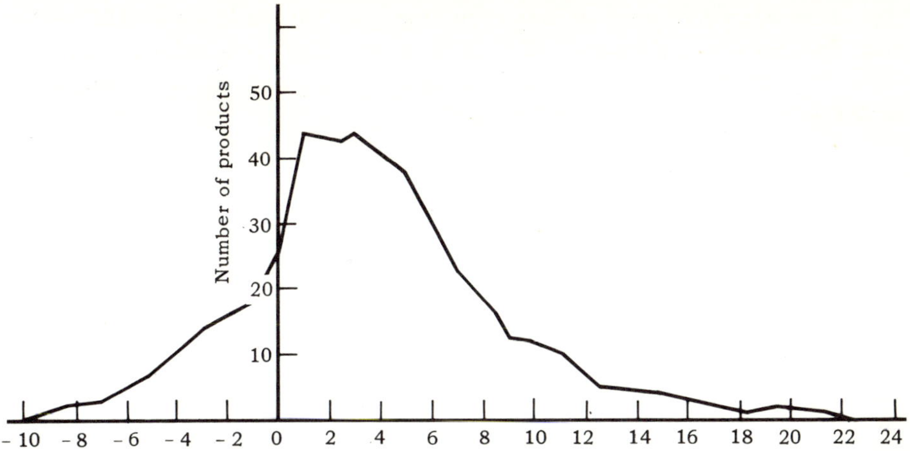

Source: See Appendix A.

the frequency distribution of tariff differences shown in Chart 2.2 indicates, there was a good deal of variation in relative UK and EEC tariffs for particular goods, and quite a few British tariffs were below their EEC counterparts, though at the same time the distribution is clearly strongly peaked, with over half of the tariff differences in the range 0 to +6, and about two-thirds in the range 0 to +8. It would thus appear that if and when the UK enters the EEC and mutual tariffs are abolished, the changes in relative UK/EEC tariffs will by no means be uniform over the 230 commodities, and adoption of the CET by Britain will similarly mean varying tariff adjustments for the commodities included. The argument holds whether or not the CET is adjusted with the entry of new members; on the other hand, it is conceivable (though not likely) that tariff reductions agreed in the Kennedy Round have served significantly to reduce the *relative* differences between UK and EEC tariffs for individual products. The point has not been taken up in the research as part of the information needed – the British post-Kennedy Round tariff – was not otherwise required in the study.

The EEC's post-Kennedy Round tariff rates, on the other hand, are used in the examination of trade diversion opportunities in Chapter 4 below, and for this reason the frequency distribution of the post-Kennedy Round CET has been inserted in Chart 2.1 (the dot-dash line). It is clear that when the Kennedy Round tariff cuts have been completed the CET will be appreciably lower than its predecessor; moreover, the degree of uniformity of rates will be considerably greater.

One point may briefly be raised at this juncture. *Prima facie* one might have expected a relationship to exist between relative UK/EEC tariffs and relative trade performances. More precisely, it might have been assumed that commodities for which the UK tariff is high in relation to the EEC's would by and large turn out to be goods in which the UK appears to be at a comparative cost disadvantage (as established by the tests described in Chapter 3), and vice versa. We carried out some analysis of this question, both for all 230 commodities taken together and by major SITC categories. The result was decidedly negative – there appears to be virtually no correlation between relative tariff rates and trade performances, either at the aggregate level or (one exception apart) at the level of SITC sections. This does not, of course, rule out the possibility of the relationship holding for individual products; some examples are pointed out in the course of Chapter 4.

3 Trade Performance and Comparative Costs: Problems and Methods

As briefly indicated at the end of Chapter 1, the approach of the present study follows that adopted in many other investigations of the industrial effects of prospective changes in trade barriers. That is to say, the main tool is an analysis of the comparative foreign trade performances of British and EEC manufacturing industry over a recent period. More specifically, the product of the statistical work – set out more fully below – is a rank list of the 230 commodities included in the study; items placed towards the top end of this list are goods in which, on the evidence of past trade performance, the UK's competitive position relative to the EEC's is strong, whilst items at the lower end of the rank order are commodities in which the UK's past record, again in relation to that of the EEC countries, is poor.[1]

The basic hypothesis is that the pattern of trade prior to the abolition of tariffs and of other trade barriers provides an indication of underlying comparative cost conditions, which in turn would be one of the determinants of the pattern of trade and hence of domestic production after the enlargement of the Common Market. *Prima facie* this appears to be a plausible hypothesis, but nonetheless it is advisable to consider the proposition set out with some care.

Two points arise. (a) There are a number of important factors which may in practice weaken the relationship between relative trade performances as observed and underlying comparative cost conditions; these are dealt with a little further below. (b) Even if it is assumed that the kind of rank order drawn up does constitute a pretty faithful reflection of comparative cost conditions, the question remains what conclusions can be drawn from it. This point is conveniently taken first.

The ultimate aim of any research into the effects of UK accession to the EEC upon British (and EEC) manufacturing industry is to predict which industries will raise their output above what it would be in the no-entry case, and by how much, and similarly to single out the sectors whose output is likely to be adversely affected, again with an indication of the extent of any such contraction. With reference to this objective a comparative cost rank order does not get one very far, because a number of other determinants (many of them presenting very serious statistical problems) are also involved in a major way. The following call for special comment.

In the first place, the effects of EEC entry on a particular British (or EEC) manufacturing sector will also depend on the size of the markets opened up by

1 Strictly speaking there are two rank orders which are then partially combined, but this refinement is of no importance in the present context.

the removal of trade barriers. For example, the British producers of an item in which the UK appears to be very competitive may find that the scope for raising exports to the EEC is relatively limited, because the proportion of total EEC demand which they supplied even before the removal of the EEC tariff was quite high, whilst in the case of a second product – in which the UK trade record has been somewhat less good, though the item is still fairly high up on the rank order – the EEC market may as yet be largely unexploited, and the opportunities for raising exports to the EEC (either at the expense of EEC producers, or of third country suppliers, or both) may therefore be excellent. Corresponding arguments apply to the growth of UK imports.[2]

In general, in order to establish the size of the effects of the EEC's enlargement upon British and EEC manufacturing industry it is necessary to go beyond an examination of comparative costs and to take into account the proportions of both the British and the EEC market which before British entry were supplied by UK, EEC and third country producers respectively. For two of the elements involved – trade between the UK and the EEC, and imports from third countries – it is easy enough to obtain the necessary data, but the third and generally no doubt the most important factor – domestic UK and EEC output for the home market – presents very considerable statistical difficulties, especially at the level of disaggregation employed in this study; in view of the limited resources at our disposal we did not even attempt to tackle the problem.[3]

Secondly, the extent to which a particular British or EEC industry can expect to raise exports to the partner above what they would otherwise have been will also depend on the reaction of local and third country producers to the relative cheapening of the British (or EEC) commodity brought about by the tariff reduction – they may adjust their terms in order to meet the new competition, or alternatively prefer to switch resources into other activities. It is not easy to see how this problem could be handled satisfactorily on a general plane, i.e., as part of a statistical analysis and without intimate knowledge of the products and markets involved.

The third point concerns the balance of payments implications of the increase in trade flows. Additional UK exports of the goods in which the UK is reasonably competitive would in themselves tend to improve the UK trade balance (and worsen the EEC's), whilst the growth of imports of goods placed towards the lower end of the rank order would worsen the British payments position. The difference between the increases in exports and imports must accord with whatever balance of trade on manufactures as a whole would be compatible with the requirement that both the UK's and the EEC's overall balances of payments (vis-à-vis all countries)

2 Given the heterogeneous nature of manufactures it is likely that even in the case of a good in which the UK is very competitive UK accession to the EEC would be followed by some rise of imports, partly at the expense of domestic (i.e., British) producers, and partly at the expense of imports from third countries.

3 Some discussion of the question of imports from third countries – i.e., of trade diversion opportunities – is, however, included in the study; the methodological issues involved are dealt with right at the end of this chapter.

would be in equilibrium. If this condition is not fulfilled, the effects of a general price adjustment would have to be built into the model (as already indicated in Chapter 1) – a relative fall in UK prices bringing about an additional expansion of exports and a reduction in imports relative to the initial increase, and a rise of UK prices having the opposite effect. Changes in relative prices would also affect UK and EEC sales to and purchases from third markets, and these would have to be brought into the picture as they clearly influence both partners' balance of payments positions.

Finally, mention must be made of the time factor. In projecting the effects of EEC entry upon particular manufacturing industries it would not be sufficient simply to take past data relating to consumption, trade, etc., and then somehow to superimpose the presumed effects of EEC entry. The only satisfactory procedure would be to construct a 'no-entry case', i.e., to forecast what the pattern of consumption and trade and hence production would be like by the end of any transition period (say, the late 1970s) assuming the UK does not join the EEC, and then to consider in what way EEC entry may be expected to alter the position.[4]

In the light of these considerations it is clear that a complete projection of the effects of UK accession to the EEC upon British and EEC manufacturing industry requires very much more than the mere ranking of products on the basis of some rather crude criteria of competitiveness, but this is all that we have been able to do.

At the same time the considerations just put forward do not preclude the use of the ranking order for at least indicating the *direction* of the EEC effect upon the output of individual industries. Given the small proportion of overall EEC and UK consumption of manufactures at present satisfied by imports from the prospective partner area,[5] limitations in the size of the market are not likely to constitute a serious impediment to the expansion of trade, consequent upon the EEC's enlargement, in many of the 230 commodities. This in turn creates a general presumption in favour of the hypothesis that the growth of UK exports to the EEC would be concentrated on products in which the UK's trading performance has been relatively strong in the past, i.e., on items placed towards the upper end of the ranking order, whilst that of UK imports (= EEC exports) would primarily be in goods found towards the lower end of the rank list. There would, no doubt, be exceptions; a few likely cases are pointed out in Chapter 4 below.

4 As explained below, the passage of time also reduces the usefulness of relative foreign trade performances as indicators of the relative competitive positions of different industries in a free trade scheme which is yet to be created, but in this case the problem is probably a less serious one.

5 As shown in Chapter 2, both the UK and EEC obtain the bulk of their imports of manufactures from third countries. *Total* imports of manufactures, moreover, constitute only a small part of each area's consumption of manufactured goods. For the UK this is, by implication, shown by the figures referred to in Chapter 1 (pp. 4–5); for the EEC see Truman, 'The European Economic Community: Trade Creation and Trade Diversion', op. cit., Table 2.

Acceptance of this hypothesis does not mean that anything very definite can be said about the relative prospects for products placed in close proximity to one another on the rank list, for if there is not much difference between the comparative cost positions of the two sets of producers, the other factors mentioned, notably the size of the markets available, may easily result in the item occupying the lower place on the rank order experiencing the greater net rise of output (or smaller net decline).

In general the relationship between a commodity's position on the rank order and the direction of the EEC effect is best regarded as a matter of probability rather than of certainty. Items at the top should mostly be goods whose output would be favourably affected. As one moves down the rank list, the number of producers who would be able to raise output would tend to decline, whilst cases of producers being adversely affected would gradually increase in frequency. Commodities found towards the bottom of the rank order would generally be goods whose output would be lower than in the no-entry position.

It is implied that somewhere away from the extreme ends of the rank order there would be a group of activities whose output would generally show little change one way or the other. As explained later in this chapter there is some presumption that this zone would be found towards the centre of the rank list, but without a full exploration of the balance of payments implications of entry – which in turn presupposes a complete assessment of the entry effects on manufacturing industry – the problem cannot really be satisfactorily resolved.[6]

It is time now to turn to the other issue raised at the beginning of the present discussion – how far past trading patterns can be used as indicators of the relative competitive position of different manufactured products in the event of UK accession to the EEC. Six points – some of them closely interrelated – may be singled out.

(1) Foreign trade comparisons by their nature relate to a past period and therefore cannot take into account the influence of changes in the circumstances of particular industries in the period – which may be quite a lengthy one – before trade barriers are abolished. These changes can come from the demand or the supply side, or result from government policy. Particularly important is perhaps the case of an industry being in a marked disequilibrium position during the period for which trade performances are being compared. For the purposes of the present study special reference might be made in this context to the alterations in trade barriers affecting UK and especially EEC manufacturing industry during the 1960s, when both EFTA and the EEC were being established; there is no reason to believe that the process of adjustment to internal EFTA/EEC free trade conditions was complete by the time to which our data refer (1965 to 1967), and any serious imbalance between demand and capacity which occurred during

6 The direction of the effects on the producers at the extreme ends of the rank list follows from the arguments put forward in Chapter 1 (pp. 4–6).

the adjustment period would clearly leave its mark on the external trade position.[7]

(2) The procedure adopted in effect treats all traded products as independent of one another and therefore disregards relationships of a complementary character. The main practical application of this point probably arises in connection with goods which are used as inputs in the production of other products. Both in the United Kingdom and in the EEC the abolition of tariffs between the two areas will alter the prices of inputs as well as the prices at which the outputs can be sold, and this is likely to affect comparative cost relationships.

(3) The existence of tariff restrictions on the trade flows being observed is another source of error. This is particularly obvious when trade barriers are so high that no trade takes place. In rather a crude fashion it is possible to make some allowance for the influence of tariffs (see below) but it cannot be pretended that the problem is thereby solved.

(4) There is a good deal of circumstantial evidence to suggest that exporters whose domestic markets are protected by tariff barriers practice price discrimination; as a result their foreign trade performances may be rather better than their relative competitive standing would normally permit, and conclusions drawn from the foreign trade data may therefore be misleading.

(5) Closely allied to the last two points is the influence of institutional factors other than tariffs – the operations of monopolies, preferential government procurement policies, differential tax systems, and such like – which serve to distort trade patterns. As already indicated (p.2, fn.2), the precise impact of these factors is difficult to evaluate and therefore to take account of. On the other hand, in so far as the conditions referred to continue to exist even after the abolition of tariffs,[8] the pattern of specialization predicted by past foreign trade performances may not be so inaccurate – though it would then fail to reflect underlying comparative cost positions.

(6) The procedure employed disregards at least two factors arising in connection with trade with third countries. (a) The adoption by Britain of the Common External Tariff (or by Britain and the EEC of a modified CET) will clearly affect the extent to which British (and possibly EEC) industries are protected against third countries. This change would influence not only output prices but also input prices; in either case there may be repercussions upon an industry's prospects in an enlarged EEC. (b) In the particular case under discussion – the UK

7 At first sight it may be tempting to argue that the November 1967 devaluation of sterling would be likely to be another 'historical' factor seriously impairing the usefulness of examining past trade performances. However, there is no obvious reason for expecting the parity change to have altered the *relative* standing of different commodities in any very substantial way, though no doubt there would be exceptions (which in themselves could be quite instructive). At the time the statistical analysis was carried out later data were not available, and the issue raised therefore had to be left unresolved.

8 Both in the EEC and in the EFTA many of the problems remain unresolved. See H. H. Liesner, 'Policy Harmonization in the EEC and EFTA', in G. R. Denton *ed., Economic Integration in Europe*, London, 1969. Further details relating to the EEC can be found in D. Swann, *The Economics of the Common Market*, Penguin Modern Economics, London, 1970, especially Chapter 3.

and the EEC – it is important to remember that the enlargement of the Communities is on balance unlikely to stop with the UK, and that at least some of the other countries now comprising EFTA will also join. The EFTA effect on the UK has in principle already been experienced, but EFTA accession to the EEC could have various consequences for the structure of production in the EEC which in turn could affect British industry. (It is possible, for instance, that at present the UK exports clothing to the EEC made from cheap cotton cloth imported from Portugal, an EFTA member; if and when Portugal joins the EEC, continental producers of clothing will also have access to cheap raw materials and therefore compete more effectively with Britain.)

Points (1) to (6) add up, at the very least, to a strong case against unqualified acceptance of foreign trade performances as reliable indicators of comparative cost conditions, and some readers may feel that the exercise is not worth undertaking at all, particularly as the results can in any case be of only limited use. Two arguments may be put forward in defence of proceeding despite the objections. In the first place, as indicated in Chapter 1 (pp. 6–8) the alternative methods of predicting the pattern of specialization in an enlarged EEC have their own serious drawbacks.[9] Secondly, it must be emphasized that many of the factors mentioned under points (1) to (6) above will to a greater or lesser extent apply to all manufacturing industries and that the position of manufactures *relative to one another* (passenger cars in relation to optical instruments or paints or cement, etc.) should therefore be less strongly affected; as the purpose of the statistical work is to find out something about the relative rather than the absolute standing of different manufacturing industries, the weight of the objections is reduced. All the same, it is clear that the results obtained can at best be interpreted in only a rather broad manner and should be supplemented by other evidence wherever possible.[10]

It is time now to give a rather fuller account of the method adopted in the present enquiry. A comparison of the foreign trade performances of the UK and the EEC for the purposes of this study could either relate to each area's exports to the other ('mutual trade analysis'), or to the exports of each area to neutral markets ('third country export analysis'). The advantage of the mutual trade analysis is that it not only takes account of variations in cost conditions as between different industries in the two areas, but that the outcome will also be affected by the pattern of demand for manufactures in the prospective partners. This means that the results of an examination of mutual trade are at least consistent with the requirement spelt out above that the availability of markets should

9 Moreover, some of the difficulties just discussed also arise in connection with the alternatives.

10 It is appropriate to mention at this point that at an early stage in the research we gave some attention to the question how far the methods used in the study would have yielded reliable predictions about the result of establishing the EEC itself. More specifically, we examined German trade performance in the period 1956/58 and compared it with her trade pattern vis-à-vis the EEC countries in 1965/67. Only preliminary results were obtained, owing to shortage of time and lack of appropriate statistics, but these seemed reasonably encouraging.

be brought into the argument, though the problem is not, of course, thereby solved. On the other hand, mutual trade analysis suffers from the disadvantage that it is in practice not possible adequately to allow for the effect of differences in trade barriers between the UK and the EEC (see below).

Third country export analysis, by way of contrast, is deficient on the demand side – there are likely to be only rather indirect links between demand patterns in the UK and the EEC and their exports to third countries – but frequently escapes the problem of differing trade barriers, though it would certainly not be true to say that there are no differences between the trade restrictions which British and EEC exports face in the rest of the world. (The imperial preference system, the association of overseas territories with the EEC, and the practice of 'tied aid' may be cited as three fairly obvious sources of differential third country trade barriers.) In view of their respective drawbacks there are no obvious reasons for preferring one approach to the other, and therefore both have been attempted below, though as will be explained, somewhat greater weight is given to the mutual trade analysis.

Foreign trade comparisons can take the form of investigating the levels of trade over a given period of time, or the changes of trade between one period and another. On balance the former alternative was preferred.[11] Analysis of relative changes over time is liable to result in misleading conclusions, because trade performances in the base period may have been very unequal, and although there are ways of dealing with this problem, there was the practical point that serious statistical difficulties would have arisen had we attempted to go back more than a few years – commodity trade figures as finely broken down as those used are not readily available before the early 1960's.

The first part of the statistical analysis concentrates on trade in 230 manufactures between the UK and the EEC over the three year period 1965 to 1967. The basic hypothesis at this stage of the argument is that if British exports of a given product, in relation to imports of that same good, have over that period 'done better' than British exports of another product in relation to British imports of the second commodity, then the British producers of the first good will enjoy a stronger comparative cost position in an enlarged Common Market than the producers of the second good. In other words, if product j is below i on the rank order, the effects of EEC entry on j are more likely to be adverse than those on i.

Exports of a particular good can be related to imports of that good in a number of ways; the method chosen for the purposes of the present enquiry is the very simple and straightforward one of taking the absolute difference between the two – i.e., the balance of trade in the commodity concerned. However, the crude rank order of the 230 product groups by value of trade balance would be strongly influenced by the fact that the 230 items vary very substantially in size of trade between the UK and the EEC. This means that the trade balance (assumed to be positive) for commodity j may be lower in the scale than that for i (exports of which also exceed imports) not because British producers of j are in a less strong competitive position vis-à-vis their EEC rivals than the producers of i, as the

11 In this respect the approach of the present study differs from that of its precursors.

rank order would seem to indicate, but simply because i is quantitatively more important than j – the market for i is larger than that for j. *Mutatis mutandis* both k and l may be relatively weak industries for the UK; they are a long way down in the ranking order, which in practice means that imports exceed exports. However, the negative trade balance for l may exceed that for k not because l producers are even weaker than k producers, but because the market for l is larger than that for k.

In order to allow for variations in the quantitative importance of different items the trade balances were normalized by dividing each balance by the sum of total EEC-cum-UK intra-trade (i.e., total imports by all the EEC countries and the UK from EEC plus UK) in that product. The rank order index for commodity i is thus found by

$$\frac{M_{ei} - M_{ui}}{\sum M_{eui}}$$

where M_{ei} *refers* to EEC imports of i from the UK (standing for UK exports of i to the EEC), M_{ui} to UK imports from the EEC, and ΣM_{eui} to total EEC-cum-UK intra-trade in i. It is clear that this procedure is very much a second-best method First of all, total EEC-cum-UK intra-trade is likely to be an imperfect proxy for size of market. Secondly, we are really dealing with two markets – the EEC market and the UK market; the procedure adopted implicitly assumes that these form a constant proportion to each other, but this is clearly not necessarily the case. It is thus apparent that the use of EEC-cum-UK intra-trade as a normalizing factor 'sweeps under the carpet' a number of problems, but in the absence of consumption and production statistics for the 230 trade items this seems unavoidable.

In the light of these considerations it may be wondered why the relationship between exports and imports was not expressed as an export/import ratio, especially as this procedure has been adopted in other investigations.[12] It is true that the export/import ratio automatically takes account of variations in the importance of different commodity groups, and therefore no normalization procedure is necessary. On the other hand, the export/import ratio fails to indicate whether or not the small size of mutual UK/EEC trade in a particular item is accompanied by a similarly low value of trade among the EEC countries, but this point is of considerable significance for present purposes. For instance, it appears important to distinguish between two goods g and h, both 'strong' goods for the UK (i.e., UK exports to the EEC exceed imports from the EEC by substantial margins) and both relatively insignificant in mutual trade, but in the case of good g trade *among* the EEC countries is also relatively insignificant, suggesting that the market is small, whereas in the case of good h there is substantial internal trade in the EEC, so that UK exports to the EEC in relation to internal trade are below average. The straightforward interpretation of the situation is that the producers of h are in a less strong position vis-à-vis the EEC market than the producers of g (provided the influence of trade barriers has been allowed for), but the export/

12 See Liesner, Balassa, and Wells (fns.16 and 17 of Chapter 1).

import ratio fails to point this out. The difficulty could only be met by dividing the export/import ratio by an indicator of market size, such as total EEC-cum-UK intra-trade, with the result that ultimately nothing is gained in relation to the trade balance approach, and something is lost because the index numbers obtained after dividing the ratios by EEC-cum-UK intra-trade are much less easily interpretable.[13]

A major objection against the analysis of mutual trade data as described so far is its failure to take account of trade barriers. Clearly, British exports of *i* to the EEC, in relation to imports of *i* from the EEC, may be more favourable than the equivalent trade balance for *j*, because the UK tariff on *i* is much higher than that on *j* whereas, say, the opposite is the case in the Common Market. A favourable trade balance for *i* in relation to that for *j* may thus be the outcome of the existing tariff structures rather than indicating differences in comparative costs. In other words, the rank order should be drawn up on the basis of trade data from which the influence of tariffs has in principle been eliminated.

This means that an attempt has to be made to estimate what additional trade flows (ΔM) there might have been between the UK and the EEC had tariff barriers not been in existence. The method used for deriving such estimates was very simple and crude, the basic formula being

$$\Delta M = d \frac{t}{1 + t} M$$

where d is a 'response coefficient', t the tariff rate and M the trade flow as recorded.

The major problem is this formulation concerns d, the response coefficient. First of all, it is important to explain what d is meant to refer to. It is evident that d should not be thought of as indicating the whole of the response of UK/EEC trade to the abolition of mutual tariffs, because the sum total of the additional trade flows would consist of both trade creation and trade diversion effects, and the latter are not relevant to an assessment of comparative cost conditions. In other words, d should only relate to the growth of trade due to trade creation.

The second problem concerns the size of d. It would appear very probable that the rate at which trade would have expanded as a result of trade creation would have varied greatly as between different commodities, even if there had been no differences in tariff barriers, but there appeared to be no way in which even approximate estimates could have been derived for the 230 items included. The only solution available was therefore to use the same coefficient for all goods.

It will be evident that the problem just referred to is in essence the same as that discussed at the beginning of the present chapter, when the output effects of EEC entry were being considered; in the light of the comments then made (and especially those relating to market size) it will be clear just how crude it is to assume the same d for all products. However, even if this objection is disregarded

13 At an early stage in the research a number of methods based on export/import ratios were tried and the results compared with the trade balance approach. In general the rank orders obtained were very similar.

there remains the problem what figure the average trade creation response should be put at. It turns out that the only solution available is to make a pretty arbitrary choice. In the literature a number of estimates are available of demand elasticities relating to trade in manufactures, with 3 to 4 perhaps the most common figures, though how far these would be appropriate in the present context is difficult to say.[14] Account must also be taken of the finding by Balassa[15] and Truman[16] that in the course of the formation of the EEC trade creation effects have greatly outweighed trade diversion effects; in other words, the increase in the growth of internal EEC trade attributed to the formation of the Common Market predominantly consisted of trade creation.

In the light of this evidence we formed the view that 3 might be a reasonable central estimate for d, and this is the figure used in the calculations. (We also conducted some experiments with $d = 5$; the results suggested that the rank order is fairly insensitive with respect to the size of d.)

The derivation of hypothetical additions to UK/EEC trade by means of the formula referred to above calls for two further comments.

(a) The recorded value of trade (M) is itself influenced by the existence of tariff barriers (as well as of other trade restrictions); the bias this imparts to the results is seen most clearly in the extreme case of t being so high that $M = O$, so that ΔM also works out at nil – clearly a nonsensical result. The fact that the bias applies to both sides of any trade balance based on hypothetical trade flows probably reduces its significance, but clearly does not mean that it can be dismissed entirely.

(b) Convenience strongly argues in favour of using as EEC tariff rates the customs duties prescribed in the Common External Tariff, despite the fact that the six Common Market countries – or, more precisely, the four customs areas – did not fully adopt the CET until mid-1968, i.e., until after the end of our period. However, as explained in Appendix A, this appears to be a reasonably defensible short-cut.

If the trade creation effects are estimated in the way described above, the adjusted trade balance becomes

$$M_{ei} + d\frac{t_{ei}}{1 + t_{ei}}M_{ei} - \left(M_{ui} + d\frac{t_{ui}}{1 + t_{ui}}M_{ui}\right)$$

or, in short,

$$M_{ei} + \Delta M_{ei} - (M_{ui} + \Delta M_{ui})$$

14 For some of the relevant considerations, and a guide to the literature, see B. Balassa *et al.*, *Studies in Trade Liberalization: Problems and Prospects for the Industrial Countries*, Baltimore, 1967, Appendix to Chapter 1.

15 B. Balassa, 'Trade Creation and Trade Diversion in the European Common Market', op. cit.

16 E. M. Truman, 'The European Economic Community: Trade Creation and Trade Diversion', op. cit.

where t_{ei} is the EEC's third country tariff on good i and t_{ui} the equivalent tariff rate for the UK. With corresponding additions to intra-trade in the denominator the ranking index for each product then works out at

$$\frac{M_{ei} + \Delta M_{ei} - (M_{ui} + \Delta M_{ui})}{\Sigma M_{eui} + \Delta M_{ei} + \Delta M_{ui}}$$

The method used in the third country export analysis can be explained very quickly. The ratio of UK neutral market exports (X_u) to EEC neutral market exports (X_e) was formed for each commodity, and these ratios were then ranked, the underlying hypothesis being that the larger are the UK's sales of a given product in third countries, in relation to the EEC's exports of the same product to the same market, the better is the UK's comparative cost position in that commodity. In other words, if

$$\frac{X_{ui}}{X_{ei}} > \frac{X_{uj}}{X_{ej}}$$

British firms producing i are assumed to occupy a stronger position in relation to EEC producers of i than do British producers of j in relation to EEC manufacturers of that commodity. No problems arise over differences in the quantitative significance of products, because these are taken care of by the use of ratios; third country exports are defined as in Chapter 2, i.e., they are equal to each partner's exports to the world *minus* exports to OECD (Europe).

This concludes the first stage of the methodological discussion. The present is thus a convenient point in the argument to return for a moment to the issues raised at the beginning of the chapter. A comparative cost ranking of the 230 commodities is of necessity a relative exercise and in itself gives no indication of the direction of the effect of EEC entry upon particular items. More concretely, the finding that British producers of i are in a stronger competitive position vis-à-vis their EEC rivals than are British producers of j, who in turn are better placed than those of k, whilst a useful piece of information, does not of itself tell one whether the level of UK output of i, j and k will rise or decline (relative to the no-entry course of events) as a result of the Community's enlargement.

As was explained earlier it is in principle not possible to find the direction in which the output of individual goods would change without working out the full quantitative effects of UK accession to the EEC upon each part of the manufacturing sector.. At the same time it was suggested that the higher is the position of a good on the rank order, the more likely is it that the British producers concerned will be in a position to raise output, and vice versa with regard to commodities found towards the lower end of the list.

There remains the question where on the rank order one would find the cluster of activities which on balance, and exceptions apart, would not be very much affected one way or the other. No definite answer can be provided, but there are two lines of argument in support of the presumption that this group of products is likely to be placed somewhere in the centre of the rank order. (a) UK/EEC trade in all manufactures (and in the 230 products considered in detail below) was, very broadly speaking, in balance, and the number of commodities in which the

UK had a surplus was again very roughly equal to that in which the balance was negative. Provided that other balance of payments effects are not such as to call for any great changes in the balance of trade in manufactures, these circumstances are at any rate in line with the stated presumption. (b) Past EEC and EFTA experience has been that whilst freer regional trade appears to have led to a very considerable expansion of trade among the respective member countries, this has not been accompanied by massive shifts of resources out of declining and into expanding sectors, in the sense of entire industries closing down (and others absorbing the factors set free). In the absence of evidence to the contrary, it may be expected that events would take a similar course in the UK/EEC case, and this in turn would reduce the chances of the group of 'neutral' activities being found outside, say, the central one-third of all products. For instance if the 'neutral' zone lay in the vicinity of commodities No. 160 to No. 200, it would be implied that the value of the additional imports of the relatively small number of items below this zone would be equal to the growth of exports of the much larger number of items higher up on the rank order – assuming again that no great change is called for in the balance of trade in manufactures as a whole. Such a sharp rise in imports of a relatively small number of items would *prima facie* be associated with abrupt cut-backs in domestic production and thus lead to a pattern of events at variance with past experience.

It is not difficult to think of objections against (a) and (b), especially perhaps the former, and the presumption that the 'neutral' zone would lie near the centre of the rank order must thus remain a weak one. But it is difficult to see how the argument could be strengthened, given the absence of further information.

Before the results of the first stage of the research are introduced two comments must be made on the data employed.

(1) Mutual UK/EEC trade, as seen from the UK, would normally be recorded on two valuation bases – exports to the EEC being measured f.o.b.. and imports from the EEC being valued c.i.f. In our case, however, the use of matrices in the collection and computer processing of data carried with it the need to keep to a consistent definition, i.e., to measure all mutual trade either f.o.b. or c.i.f. Because of the incorporation of tariffs in the analysis the case in favour of taking c.i.f. data seemed on balance slightly stronger, and hence all mutual trade figures used in the detailed computations are c.i.f. values.[17] In any event, sample checks suggested that a complete substitution of f.o.b. for c.i.f. data would not have made any significant difference to the results. UK and EEC third country exports are, of course, measured f.o.b.

(2) The second comment deals with the far more important matter of the commodity classification employed. It was clear right from the beginning that there are strong arguments in favour of using a detailed commodity breakdown for the kind of analysis attempted below, far more disaggregated than that found in earlier studies. In the first place, the aim of the research is, of course, to assess the comparative position of different British products in a wider Common Market,

17 As readers may have noticed, most of the discussion is nonetheless worded in terms of UK exports to the EEC and EEC exports to the UK – simply because it turned out to be rather easier to conduct the argument in this fashion.

the underlying assumption being that if and when tariff barriers are removed there will be a shift in the pattern of production in favour of strongly placed goods, whilst the output of comparatively weak products would decline, relatively if not absolutely. The more aggregated the product groups used in the investigation are, the greater is the danger that a substantial proportion of the shift will occur between the goods covered by one and the same heading and thus not be predicted by the procedure adopted in the study.[18]

Secondly, both the EEC and the UK classify their tariffs with reference to the Brussels Tariff Nomenclature (BTN) whereas the trade statistics are arranged according to the Standard International Trade Classification (SITC). Moreover, the number of separate tariff headings greatly exceeds the number of items even in the finest SITC classification available. In consequence the tariff rates assigned to individual product groups are subject to some error; the less detailed the commodity breakdown used, the greater this error is likely to be.

The issue of commodity disaggregation is closely linked to the question of coverage. Most of the 230 products are listed in the SITC as four-digit items (three-digit data were used only when the SITC does not provide a finer classification). The disadvantage of using the four-digit data lies in the reduction in the coverage of manufactures included, because a complete breakdown of trade by four-digit items was not available. Coverage was further reduced because of the exclusion of a few products to which the theoretical model underlying the analysis does not appear to be applicable. Trade in newspapers and periodicals (SITC 892.2), for example, is unlikely to be a reliable indicator of the relative competitive positions of the respective domestic producers of these goods. However, only a few products were left out on these grounds, the guiding principle being 'when in doubt, include'.

The extent of the final coverage of trade in manufactures, broken down by main SITC sections, is shown in Table 3.1. The figures (which relate to intra-trade) show that despite the use of four-digit SITC data and the omission of a few items the value of excluded commodities does not significantly exceed 15 per cent in any major category; measured overall the coverage comes to 93 per cent.

The decision to work with as many as 230 commodity headings made it necessary to divide the data into reasonably manageable units. Accordingly the presentation and discussion of the results of the work will largely be conducted with reference to nine industries or industry groupings. However, it must be pointed out that any division of SITC items into industries is of necessity rough and ready, because the main criterion for the SITC is the nature of the good, and the product classification which results frequently cuts across conventional industrial boundaries (usually drawn in terms of the origin of the goods concerned). The need to work with units of a size reasonably well suited to the particular presentation adopted further added to the difficulty of obtaining clear-cut groupings.

The list of industries finally chosen is set out in Table 3.2, together with a definition of each group in terms of the SITC items included. A full description of these items would run over a large number of pages and the reader must

18 See Chapter 5 for further discussion of the issue.

therefore to be referred to the official sources,[19] but a short summary is provided, for the appropriate goods, at the beginning of each industry section in Chapter 4 below.

Table 3.1 *Coverage of UK and EEC Trade in Manufactures*

(measured with reference to intra-trade, 1965/67 annual average)

SITC Description	Total intra-trade ($m., c.i.f.) (1)	Intra-trade in products included in the study ($m., c.i.f.) (2)	(2)/(1) (per cent) (3)
5 Chemicals	2,521	2,210	87.7
6 Manufactures classified chiefly by material	7,563	6,935	91.7
7 Machinery and transport equipment	7,878	7,762	98.5
8 Miscellaneous manufactured articles	2,598	2,198	84.6
5-8 Total	20,560	19,105	92.9

Source: Col. 1: As for Table 2.1
Col. 2: Tables 4.A.3, 4.B.3, 4.C.3, etc.

Table 3.2 *An Industrial Grouping of SITC Sections* 5-8

Industry	Composition	
	SITC[a]	Number of products
A. Chemicals	5	38
B. Manufactures of leather, rubber, wood, paper, and non-metallic minerals	61, 62, 63, 64, 66, 821, 831, 841.3, 851	31
C. Textiles and clothing	65, 84(ex. 841.3)	31
D. Iron and steel	67	24
E. Non-ferrous metals and miscellaneous metal manufactures	68, 69	24
F. Mechanical engineering	71	29
G. Electrical engineering	72, 812.4, 891.1, 891.2	19
H. Transport equipment	73	14
I. Miscellaneous manufactures	86, 891.8, 892.9, 893, 894.2, 894.4, 895.2, 897.1	20

a Full details of the composition of each industry in terms of the SITC groups/subgroups included can be found in Ch. 4 (Tables 4.A.1, 4.B.1, 4.C.1, etc).

19 *Standard International Trade Classification, Revised*, United Nations, Department of Economic and Statistical Affairs (Statistical Papers, Series M, No.34), New York, 1961.

It was stated earlier in this chapter that the first objective of the statistical work was to rank the 230 products on the basis of a set of criteria relating to mutual UK/EEC exports in the period 1965/67, rough adjustments having been made for the influence of tariffs. As repeatedly emphasized the ranking order so derived is pretty rough and ready, and too much cannot be read into the precise position of a particular product, especially with reference to the direction of the EEC effect on output. For this reason the rank order has been divided into five classes of products each containing 46 commodities, and within each group goods are arranged by SITC classification rather than by rank number; the rank order has in effect been reduced to five steps. Class I contains commodities with ranking numbers 1 to 46 and thus represents the group of products which, from the point of view of the UK, enjoy the best trade record; given the interpretation of the rank list put forward earlier (pp. 19–20 and p. 27) the direction of the EEC effect on the producers of most of these items should be favourable. The products in class II (nos. 47–92) seem also likely on the whole to be quite strong performers. Class III refers to goods assigned to the centre of the rank order (i.e., nos. 93–138) and therefore to products with a middling record; as pointed out earlier, there is a weak presumption that a relatively large number of the items in this class will be comparatively little affected one way or the other. Classes IV and V (nos. 139–184 and 185–230 respectively) contain commodities which, in the light of the evidence, must be described as likely to be relatively poor performers, particularly, of course, those in class V. *Mutatis mutandis* these are the products with the best (class V) and good (class IV) record and prospects from the point of view of the EEC.

It is clear that the procedure just discussed is open to at least one serious objection. Grouping the commodities into five classes if anything accentuates the significance of the precise ranking place of a product at the cut-off points – between nos. 46 and 47, 92 and 93, etc. In order to deal with this problem we decided to draw specific attention to the eight products placed around each of the cut-off points, marking the four goods which 'just made' the higher class with an asterisk and those at the top of the next lower class with a dagger. (The decision to make the number of goods to be singled out in this way eight, rather than, say, six or ten or some other number simply reflects our choice as between two conflicting aims – to play down the precise ranking order derived, and to avoid having to deal with a large number of special cases.)

It is appropriate to ask how the five classes compare in terms of the value of total intra-trade, the yardstick used earlier to assess the quantitative significance of the items covered. The answer is shown in column 1 of Table 3.3 (which otherwise is not relevant to the present discussion). Class III is clearly much the largest class, and whilst class V is the smallest, it also emerges that classes IV and V taken together are of a similar size to the combined classes I and II. In terms of absolute value of trade the four classes grouped around the centre can thus be said to be in broad balance. This adds some support to the hypothesis that the 'neutral' zone (i.e., the area between activities which would mostly expand and those which would generally contract) would lie near the centre of the rank order (see pp. 27–28 above) but is certainly not sufficient to clinch the argument.

Table 3.3 *Intra-trade and Exports to Third Countries by Performance Class*, 1965/67

(Mutual Trade Ranking)

Class (mutual trade ranking)	Intra-trade ($m., c.i.f.) (1)	UK exports to third countries ($m., f.o.b.) (2)	EEC exports to third countries ($m., f.o.b.) (3)	(2)/(3) (per cent) (4)
I	3,593	2,047	2,033	100.7
II	3,771	1,501	2,923	51.4
III	5,183	1,490	3,931	37.9
IV	3,995	811	2,382	34.1
V	2,563	662	2,401	27.6
Total	19,105	6,511	13,670	47.6

Notes: (i) Source: Tables 4.A.2, 4.B.2, 4.C.2, etc.
(ii) The classing of commodities with reference to mutual trade is explained on p. 31.

The results of the mutual trade analysis arranged in the manner indicated are set out in Appendix Table B.1. There is no point in considering the table in any detail at the present stage, because the discussion would inevitably overlap with the material of the industry sections in Chapter 4.[20] The only aspect to be taken immediately is the comparison of the results so far obtained with the evidence provided by the second approach – the examination of relative UK and EEC exports to third countries. As indicated earlier, we decided at an early stage of the work to consider not only mutual UK/EEC trade as an indicator of the comparative cost positions of the 230 commodities included, but also relative export performances in neutral markets. A second rank order, based on the alternative set of criteria, was thus drawn up. This rank list was treated in the same way as the first one – i.e., divided into five classes of descending order of competitiveness from the UK point of view, with each class containing 46 commodities. A table showing the composition of the five classes is shown in the Appendix (Table B.2).

The results of the two methods can be compared in a number of ways. One possibility is simply to examine total UK and EEC third country exports for each performance class derived from the mutual trade ranking. The figures are set out on the right hand side of Table 3.3. As column 4 of the table indicates, there is a consistent relationship between mutual trade and third country export performances, in the sense that the lower the class in the mutual trade ranking, the worse also the relative performance of the group of commodities concerned in neutral markets.

The first impression is therefore that the two approaches lead to broadly similar results. This conclusion derives a measure of support from the comparison of the two rank orders as set out in Table 3.4. The cells lying in the diagonal band picked out by the solid lines refer to the products for which the two tests give the same answer; whether one looks along the columns or along the rows, the figure in this cell is nearly always higher than any other in that column or row.

20 It is for this reason that the table has been relegated to the Appendix.

Table 3.4 *A Comparison of Rank Orders by Class*

Mutual trade ranking \ Third country export ranking	I	II	III	IV	V	Total
I	21	14	7	3	1	46
II	10	14	4	11	7	46
III	8	8	17	8	5	46
IV	2	8	13	13	10	46
V	5	2	5	11	23	46
Total	46	46	46	46	46	230

Source: Appendix Tables B.1 and B.2

Thus, if according to one rank order product *i* is placed in, say, class III, the probability of *i* being put into the same class by the second rank order is generally greater than that of *i*'s assignment to any of the remaining classes. On the other hand, apart from one case – the agreed class V cell – the number of commodities in the diagonal cells is less than half the total number of goods in each class (46), and in two cases (agreed classes II and IV) it is as low as 14 and 13 respectively (less than one-third of the total number). The similarity between the two rank orders can thus not be pressed too far. This is confirmed by Spearman's correlation coefficient between the two rank orders, which works out at 0.53 (statistically significant at a 1 per cent level).

The partial conflict of evidence raises the awkward question in what way if any the results of the two approaches should be combined. One solution would be to form some kind of average of the two ranking orders so as to derive one single indicator. This method we rejected as unsound, because it would be implied that differences between the rank orders are of no significance. Instead we ultimately decided to base the detailed analysis of trade performances initially on only one class list (that derived from mutual trade data) but subsequently in effect to separate out those items – termed conflicting evidence goods – which the second approach (the third country export analysis) placed two or more classes below or above the mutual trade performance class. As Table 3.4 indicates, the number of products involved (those placed outside the broken diagonal lines) is 64, 28 per cent of the total.

It would appear that the criterion employed for distinguishing conflicting evidence cases has at least one unsatisfactory feature, given the dispersion of products *within* each of the performance classes.[21] Some thought was therefore given to the possibility of designating as conflicting evidence products those items for which the two rank orders differed by more than a given number of steps – the obvious choice would be 91 – rather than by more than one class. There

21 Assume that product *i* is placed near the top of class II by the mutual trade test and near the bottom of class III by the third country export ranking, whilst *j*'s places in the two rank orders are at the bottom of class II and the top of class IV respectively. *j* will be singled out as a conflicting evidence good and *i* not, despite the fact that in terms of rank order steps the difference between the placings is greater in the case of *i* than it is in that of *j*.

were arguments in favour of either criterion, but on balance the class difference seemed the better option. In any case, the number of products which would have ceased to be conflicting evidence items had we gone over to the alternative approach is not large (23); a list identifying the products concerned is provided in the Appendix (see Table B.3).

It is implied that whenever the difference between the performance rankings is only one class the third country export result is disregarded. Two sets of considerations prompted us to take this line. In the first place, it appeared to us that for the purposes of the present study somewhat greater weight should be attached to the arguments in favour of the mutual trade analysis than to those supporting the third country export approach. In marginal cases we would thus tend to 'put our money' on the mutual trade ranking.[22] Secondly, it so happens that the 64 conflicting evidence products tend to be goods which are relatively unimportant in third country trade, in the sense that both UK and EEC third country sales of the items concerned ($20.8 million and $54.7 million respectively) are on average somewhat below the corresponding figures for the 230 items taken together ($28.3 million and $59.4 million respectively). This point is not taken into account in comparisons of rank orders as attempted in Table 3.4, or in rank correlations of the Spearman type (though it does affect the outcome of an examination of third country exports by mutual trade performance – the object of Table 3.3). It can thus be argued that the divergencies between the two rank orders are in a sense less marked than Table 3.4 and a rank correlation coefficient of 0.53 would appear to indicate; this in turn weakens the objections against putting particular emphasis on one of the two sets of results. At the same time it must be added that the implications of variations in the relative size of third country exports have not been fully explored, and it is thus quite possible that further analysis and thought would have led us to rather different conclusions.

It was stated at the beginning of this chapter that the overall effect of UK accession to the EEC upon any particular British or EEC industry will depend not only upon that industry's competitive position vis-a-vis its rivals in the prospective partner, but also upon a number of other factors. One of these is trade diversion, i.e., the possibility of replacing imports from third countries (still subject to tariff restrictions) in the partner's market. The final section of the present chapter sets out how far the question of trade diversion can be tackled within the framework of this study.

The degree to which a particular industry may expect to enhance its sales in the prospective partner country through trade diversion will be determined by several factors; for present purposes the following four require mention. (The argument is conducted with reference to a British industry; it applies *mutatis mutandis* to manufacturers in the EEC.)

22 The line adopted is, of course, a matter of judgment, and in the light of the difficulty of making proper allowance for the influence of tariffs (see pp. 25–27), it may be suggested that our judgment was wrong. However, the point made earlier – that the mutual trade analysis takes some account of the consumption side of the enlarged market – seemed to us to be of sufficient importance to sway the argument in these cases of doubt.

In the first place, there is a general presumption that the stronger is the comparative cost position of the British producers in question in the enlarged EEC, the better placed, *ceteris paribus*, they will be to benefit from trade diversion. An industry which would be in a relatively weak competitive situation in the enlarged EEC and which would therefore be expected to find its market (presumably in large measure a domestic one) being invaded by its rivals in the other EEC countries would, on the face of it, not be able to make much use of the fact that relative to producers in third countries its position in the partner country would improve. Given the heterogeneous nature of most manufactures (i.e., the possibility of increasing the sales of specialist lines), it would probably be unreasonable to suggest that such an industry could not obtain any gain from trade diversion, but for working purposes it would appear sufficient to concentrate on the products in which the UK and the EEC respectively are relatively strongly placed. The assessment of opportunities for trade diversion will therefore be confined, in the case of British industries, to products placed in performance classes I and II, and for EEC industries on commodities assigned to classes IV and V. (Class III products will be included in the analysis only if they are conflicting evidence commodities, i.e., placed by the third country export test in class I or class V.)

Secondly, the scope for trade diversion in favour of UK exporters will clearly depend on the extent to which the EEC market would be supplied by third country producers in the absence of UK accession to the EEC; the larger would be imports from third countries, the greater the possibility of trade diversion, other things being equal, and vice versa. There was no possibility of our forecasting the EEC (and UK) level of third country imports several years hence,[23] but past imports should serve as a crude indicator of the relative magnitude of the markets enjoyed by third country suppliers. Because of the probability that the majority of the European OECD countries would in one way or another follow the UK into the EEC, 'third country imports' are for these purposes best defined as imports from the rest of the world *minus* imports from OECD (Europe).[24]

The third factor is the level of the tariff prescribed for the product in question in the customs schedule of the enlarged EEC, as this determines the degree of discrimination against third country imports; the higher the tariff, the greater the scope for trade diversion, and vice versa. It is at present impossible to say in what way, if any, the existing Common External Tariff of the EEC will be

23 Any such attempt would *inter alia* have to show how British and EEC third country imports of the 230 commodities may be expected to respond to the Kennedy Round tariff cuts.

24 Both the UK and the EEC are, of course, part of OECD (Europe); mutual UK/EEC trade is thus automatically omitted. On the other hand, third country imports as defined include imports from the EEC's Associated Overseas Territories as well as from countries (predominantly in the Commonwealth) which may be granted Associated Overseas Territory status when the EEC is enlarged. This trade would not be subject to trade diversion in the sense discussed, but as little of it is likely to consist of manufactures the point can safely be disregarded.

modified if and when the UK and other EFTA countries join, and the only practical solution is to assume that the CET will continue unchanged. On the other hand, this does not mean that it would be satisfactory simply to fall back on the CET rates used elsewhere in the study (those in force in 1965/67), as the Kennedy Round reductions should be fully implemented by the time the UK would join. For this reason the duties incorporated in the examination of trade diversion will be those of the post-Kennedy Round CET.

Finally, there is the reaction on the part of the overseas supplier to the deterioration of his relative position in the EEC market. It is open to third country exporters to adjust their terms so as to protect their sales to the EEC, and the extent to which they are prepared to do this will clearly affect the incidence of trade diversion. Unfortunately there appears to be no way of predicting just what the reaction of third country suppliers will be, either in general or in any particular case, and in practice the point must therefore be disregarded. (One of the determinants would presumably be the height of the tariff; the influence of the volume of sales could work either way.)

An examination of the extent to which a British class I or class II producer may expect to raise sales through trade diversion can therefore be based on only two factors (the first of which is no doubt affected by the second) – the level of imports from third countries and the height of the Common External Tariff. However, there appears to be no satisfactory way in which they can be combined into a single indicator. Moreover, even if this difficulty could be overcome, it is not at all obvious how the result should be interpreted. In principle one would wish to compare an industry's gain (in terms of additional exports sold) from trade diversion with the growth of its trade due to specialization within the enlarged EEC, but this is a piece of evidence which, as several times mentioned above, cannot be obtained from a mere ranking of the 230 items. Examination of the scope for trade diversion must therefore be restricted to the setting out of the relevant information for classes I, II, IV and V and for conflicting evidence commodities in class III, attention being drawn to particular cases.[25]

25 The procedure adopted raises an awkward problem in connection with conflicting evidence items not assigned by either of the two tests to class III. For instance, if a product is placed in class II by the mutual trade rank order but in class V by the third country export test, it is not clear whether it is British or EEC producers who have the competitive edge and who would therefore be in the better position to exploit opportunities for trade diversion. There was therefore an argument for omitting the 39 products concerned from consideration (and thus to exclude them from the tables dealing with trade diversion – 4.A.4, 4.B.4, etc.). In the end it seemed preferable not to proceed in quite such a drastic fashion, but third country imports are in general only shown and discussed in relation to the mutual trade rank order placing. In the tables mentioned the descriptions of conflicting evidence products are printed in italics in order to draw attention to the ambiguity just referred to.

4 Trade Performance by Industry

The purpose of the present chapter is to discuss the results of the statistical work grouped in accordance with the industrial classification adopted. The chapter is therefore subdivided into nine industry sections; in each of these the discussion follows a broadly similar pattern. An introduction dealing with the composition of the industry in terms of the SITC items included leads to a brief analysis of UK and EEC tariff barriers on the products concerned as applied in 1965/67. Two aspects of tariffs are considered, each with reference to a frequency distribution chart – the two sets of tariff rates themselves, and the differences between UK and EEC tariffs for the same product.[1]

The next stage is a summary of the industry's trade record, in the aggregate and by performance class (mutual trade ranking). Particular attention is paid to two indicators – the balance of mutual UK/EEC trade in the products concerned,[2] and the ratio of UK to EEC exports to third countries (the neutral market index, expressed as a percentage). Brief reference is also made to the individual ratios of UK to EEC third country exports for the five performance classes, as this provides some indication of the degree to which the two performance measures agree with one another. (The equivalent exercise was undertaken earlier for all 230 products taken together – see p. 32 and Table 3.3)

The main part of each industry section is devoted to the details of the statistical evidence concerning the comparative cost positions of the products included. However, it should be emphasized that the discussion makes no attempt to deal with each of the 230 commodities individually. Rather, the results are left largely to speak for themselves, and comment is reserved for points which appear to be of special interest. The reasons for adopting this treatment are partly a matter of length, and partly reflect our awareness that the kind of comment which is required calls for intimate knowledge of individual products and markets, knowledge which we certainly do not possess. Within this overall approach there is naturally some tendency for the discussion to concentrate on 'important' products; in the absence of data relating to values of domestic consumption and/or production, the yardstick used for assessing the quantitative significance of

1 These differences do not play any particular role in the subsequent argument, but some readers, especially those interested in the position of a specific industry or set of products, should find the information useful.

2 These trade balances are calculated from figures in which trade in both directions is recorded c.i.f. (see p. 28) so that they show a larger surplus (or a smaller deficit) of UK exports relative to imports that one would obtain from using the British trade accounts.

individual items has been the value of intra-trade (trade *among* the EEC countries together with EEC exports to and imports from the UK).

It should perhaps also be made clear that on the whole the detailed evidence has been discussed without referring back to the overall indicators of each industry's position (described in the penultimate paragraph). Any attempt to tie up the two sets of data – to 'explain', say, a divergence between the overall trade balance and the neutral market index by reference to the trade values for individual items – would have necessitated a great deal of involved argument of ultimately rather questionable validity.

Finally, at the risk of the argument becoming repetitive one of the points made in the last chapter should again be emphasized at this stage. Even if the ranking procedure were fully reliable (which evidently it is not), the assignment of a commodity to a particular class does not carry with it any firm indication as to the direction of the effect of UK accession to the EEC upon the producers concerned. All that can be said is that there is some probability that the British producers of items placed in classes I and II (especially the former) will do well, in the sense that their output will be larger than it would be if the UK remained outside, and vice versa with regard to British producers of commodities found in class IV and, in particular, in class V.

When specific commodities are being considered it is also important to remember that individual UK (or EEC) producers of any given item will not necessarily all be affected in the same way. A particular product may, say, be assigned to class IV, and although British producers *as a whole* may therefore find the going relatively tough, there may well be individual units which are rather more favourably, or alternatively even less favourably, placed, because of supply or demand factors peculiar to the firm.

Two points should be made concerning the last topic in the industry sections – opportunities for trade diversion. (1) As indicated above (p. 35) one of the determinants of the size of any trade diversion effects is the value of third country imports before the abolition of trade barriers among the partners. The discussion will therefore concentrate on commodities distinguished by a relatively high level of third country imports into the EEC (or the UK); with the average 1965/67 value of third country imports amounting per product to $21.5 million a year ($16.4 million for the UK and $26.6 million for the EEC), 'relatively high level' means imports in the region of $20 million or above.[3] (2) As also explained earlier, the tariff data used in the examination of trade diversion opportunities are based on the EEC's prospective post-Kennedy Round customs schedule; they therefore differ from the CET rates discussed at the beginning of each industry section (in general they are, of course, lower).

3 The method used for scaling third country imports is far from satisfactory, but in the absence of estimates of the size of each partner's market supplied by domestic producers there appeared to be no preferable alternative.

A. Chemicals

The first industrial section deals with chemicals. This is the largest of the industry groupings in terms of the number of products included, though it is of only average size in terms of intra-trade. The 38 commodities separately distinguished are listed in Table 4.A.1.

Tariffs on chemicals as applied in 1965/67 are taken first. Chart 4.A.1 shows the two frequency distributions of the tariffs set out on the right hand side of the table. It appears that both UK and EEC tariffs on chemicals were rather high; this is particularly obvious with regard to the UK, in view of the second peak of rates in the 25 to 29.9 per cent range (largely referring to basic organic chemicals), but a comparison of the graph with Chart 2.1 suggests that it is also true in the case of the EEC. The frequency distribution of tariff differences (which are listed in the last column of the table) is shown in Chart 4.A.2. It is clear that there were some very substantial divergencies between UK and EEC tariffs for the same product; only 15 tariff differences fall in the –2 to +2 bracket, there are five cases of EEC tariffs exceeding the corresponding UK rates by 3.1 percentage points or more, and ten cases (over one quarter of the total) of British tariffs in excess of their EEC equivalent by at least 9 points. It might be added that, as inspection of individual tariff rates in Table 4.A.1 shows, seven out of the last-mentioned ten products are organic elements and compounds (and thus part of the basic chemicals group), whilst the goods distinguished by an EEC tariff in excess of the British by five percentage points or more are all 'processed chemicals' – medicaments, plastics (two items), detergents, and starches. (At the same time there are other processed chemicals for which UK rates exceeded EEC rates by substantial margins).

Data relating to UK and EEC trade in chemicals during 1965/67, both at an aggregate level and by performance class, are shown in Table 4.A.2. Total UK exports to the EEC averaged $223.0 million, and EEC exports to the UK $303.4 million; the crude trade balance was thus in the EEC's favour by about $80 million. Exports to third countries on average came to just under $700 million for the UK and $2,000 million for the EEC, and the overall index of neutral market performance (the ratio of UK over EEC exports) was therefore 34.8. Since for the 230 manufactures as a whole mutual UK/EEC trade was nearly in balance (the exact figure is a surplus for the UK of $175.3 million) and the neutral market export ratio came to 47.6 (see Table 3.3), it appears that taken as a whole, and before allowance for tariffs on mutual trade, chemicals do not look like one of the stronger sectors from a UK point of view.

The other aspect of the table which is worth noting is the limited extent to which the two performance measures are related to each other. As column 6 shows, there is virtually no difference between the third country export performances of the goods which the mutual trade ranking places in classes II to IV. This suggests a rather poor correlation between the two rank orders, at least as far as the commodities in these three classes are concerned. But it is satisfactory to note that the third country export ratio is much higher for class I and much lower for class V.

The detailed results are set out in Table 4.A.3; the following features call for some comment. The nine items in class I are without exception processed chemicals. Particularly noteworthy is perhaps the assignment to that class of all three constituents of SITC 533 (which comprises paints and colouring materials) and of the three representatives of SITC 54 – Medicinal and Pharmaceutical Products. However, it must be added that in this last case all three products are conflicting evidence commodities, and that one of these, Vitamins etc., is the only product out of the whole list of 230 which the mutual trade ranking places in class I and the third country export ranking in class V.

Class II contains only four items, three of them basic chemicals (and also conflicting evidence goods), and one chemical product – Explosives. It may perhaps be asked how far trade in this product is based on commercial criteria, but we are not in a position to supply an answer.

No special comment appears to be called for with regard to class III. Class IV contains the two products which quantitatively are the most important of all 38 chemicals – Condensation Products and Polymers, both constituents of the plastics group. In general, when considered with reference to mutual trade, plastics do not appear to be particularly strongly placed; the two items mentioned are assigned to class IV, a third, Cellulose and Derivatives, only makes class III, and whilst the fourth, Other Plastic Materials, is placed in class I, it is clear that this commodity is of minor significance in quantitative terms. However, the evidence of relative exports to third countries suggests a rather better position for the main plastics items; Cellulose and Derivatives, and Condensation Products, are both class II items in the alternative rank order, whilst Polymers are assigned to class III.

A clear case of poor performance, on the face of it, is the fertilizer section. Two out of the three commodities covered (Nitrogenous and Potassic Fertilizers respectively) are placed in class V, and the third, Phosphates, in class IV. It might be added that in terms of third country export performance all three are assigned to class V.

Finally, attention should be drawn to the rather pessimistic impression created by the results for basic chemicals (SITC 51). There are altogether 14 items in this section; nine of these are uncontestedly[4] placed in class IV or class V, and four of the remaining five are conflicting evidence cases, with one of the performance indicators determining an assignment to one of the two last classes. (The fourteenth, Other Metallic Salts II, only just makes class III on mutual trade, though does so comfortably with reference to third country exports). The fact that the British tariff on many of the products concerned is relatively high (see Table 4.A.1) appears to lend emphasis to the conclusion that the outlook for these products seems relatively unfavourable.

Discussion of opportunities for trade diversion is based on Table 4.A.4. This sets out the EEC's third country imports of the commodities in classes I and II (and one conflicting evidence item in class III), and the UK's third country

4 The 'uncontested' (or 'agreed') assignment of a product to a given class means that there is at most one class difference between the two rank order placings. In other words, it is not necessarily implied that both rank orders would assign the item to the same class.

imports of the products in classes IV and V, together with the relevant rates of the post-Kennedy Round Common External Tariff. The figures for the products in classes I and II (upper part of the table) show that the EEC's three leading third country imports are all conflicting evidence commodities, so that it is not clear that the UK producers concerned would in fact be in a position to exploit opportunities for trade diversion on a substantial scale. The most promising case, relatively speaking, may therefore be Other Organic Chemicals.

Looked at from the point of view of producers in the EEC (lower part of the table) the most favourable cases with respect to trade diversion would appear to be Nitrogen Compounds and Polymers, followed by Organo-inorganic Compounds, Starches, and Organic Acids. Plastic materials classified as Condensation Products would also appear in this list but for the conflict of evidence concerning the competitive position of that part of the industry. (If the neutral market export performance were regarded as the more reliable indicator the item would deserve being added to those which appear to afford good scope for gains from trade diversion for British producers, as the figure for the EEC's third country imports – $39 million – is also relatively high.)

B. Manufactures of Leather, Rubber, Wood, Paper and Non-metallic Minerals

The present section is the first of four dealing primarily with the goods which the Standard International Trade Classification groups under the catch-all heading Manufactures Classified Chiefly by Material (SITC 6).[5] However, unlike the others it contains a rather miscellaneous collection of items, as the title indicates. It would thus not be appropriate to talk in terms of a single industry.

It should also be said at the outset that a number of the commodities covered (principally in the wood and paper groups) are products in which Scandinavian members of an enlarged EEC may be presumed to occupy an especially strong position. As explained in Chapter 3 our analysis of comparative cost conditions takes no account of the possibility that other European countries may in some form or another accompany the UK into the EEC, and whilst this omission is likely to have an adverse effect on the reliability of all our results, the consequences may be particularly serious in the case of the industry group with which we are at present concerned. In other words, some of our conclusions, and particularly those based on Tables 4.B.3 and 4.B.4, may be quite erroneous.

Table 4.B.1 lists the 32 products concerned, together with the relevant UK and EEC tariffs as applied during 1965/67. The frequency distribution of tariff rates (Chart 4.B.1) is fairly similar to that for all commodities (see Chart 2.1), with EEC rates somewhat lower than British, but as in the case of chemicals the distribution of tariff differences (Chart 4.B.2) turns out to be rather uneven. There is a cluster in the 0 and +2 intervals, i.e., tariff differences are quite small for a substantial proportion of the goods included, and a second peak in the +6 and +8 ranges. Examination of Table 4.B.1 indicates that there are no clearly defined associations between either of the two clusters of tariff differences and a

5 In addition to items from SITC 6 two of the four sections also include a few products from SITC 8.

Table 4.A.1 *Chemicals: Products Covered and Tariff Rates*

SITC	Description	Tariffs (1965/67; per cent) UK	EEC	UK-EEC
512.1	Hydrocarbons and their halogenated, sulfonated, nitrated or nitrosated derivatives	29.0	15.0	14.0
512.2	Alcohols, phenols, phenol-alcohols, glycerine	22.2	12.8	9.4
512.3	Ethers, epoxides, acetals	33.3	16.8	16.5
512.4	Aldehyde-, ketone- and quinone-function compounds	29.9	16.2	13.7
512.5	Acids and their halogenated, sulphonated, nitrated or nitrosated derivatives ('Organic acids')	27.0	16.5	10.5
512.7	Nitrogen-function compounds	28.5	16.0	12.5
512.8	Organo-inorganic and heterocyclic compounds	29.7	16.3	13.4
513.2	Other chemical elements such as chlorine, other halogens, sulphur, etc.	8.2	6.5	1.7
513.3	Inorganic acids and oxygen compounds of non-metals or metalloids	16.4	10.2	6.2
513.5	Metallic oxides, of kinds principally used in paints	17.4	11.2	6.2
513.6	Other inorganic bases and metallic oxides	18.9	11.6	7.3
514.2	Other metallic salts and peroxysalts of inorganic acids (I)	15.2	12.3	2.9
514.3	Other metallic salts and peroxysalts of inorganic acids (II)	14.6	12.8	1.8
514.9	Other inorganic chemicals such as liquid air, hydrogen peroxides, phosphides, etc.	9.6	10.3	−0.7
521	Mineral tar and crude chemicals from coal, petroleum and natural gas	9.3	3.8	5.5
531.0	Synthetic organic dyestuffs, natural indigo and colour lakes	20.4	15.3	5.1
532	Dyeing and tanning extracts, and synthetic tanning materials	7.1	8.2	−1.1
533.1	Other colouring materials	20.0	13.0	7.0
533.2	Printing inks	12.5	15.0	−2.5
533.3	Prepared paints, enamels, lacquers, varnishes, etc.	13.4	13.6	−0.2
541.1	Vitamins and provitamins	25.5	13.0	12.5
541.3	Penicillin, streptomycin, tyrocidine and other antibiotics	13.4	13.6	−0.2
541.7	Medicaments	7.0	24.5	−17.5
551	Essential oils, perfume and flavour materials	10.0	9.6	0.4
553.0	Perfumery and cosmetics, dentifrices and other toilet preparations	17.3	15.0	2.3
554.1	Soaps	17.3	15.0	2.3
554.2	Surface-acting agents and washing preparations ('Detergents')	7.0	15.0	−8.0
561.1	Nitrogenous fertilizers and nitrogenous fertilizer materials	25.0	10.0	15.0
561.2	Phosphatic fertilizers and phosphatic fertilizer materials	11.0	3.0	8.0
561.3	Potassic fertilizers and potassic fertilizer materials	0	3.0	−3.0
571	Explosives and pyrotechnic products	16.2	14.2	2.0
581.1	Products of condensation, polycondensation and polyaddition ('Condensation products')	14.0	19.0	−5.0
581.2	Products of polymerization and copolymerization ('Polymers')	15.0	20.0	−5.0

Table 4.A.1 *Continued*

SITC	Description	Tariffs (1965/67; per cent) UK	EEC	UK-EEC
581.3	Regenerated cellulose, chemical derivatives of cellulose and vulcanised fibre	28.2	17.0	11.2
581.9	Other artificial resins and plastic materials	13.5	12.5	1.0
599.2	Insecticides, fungicides, etc.	10.0	9.7	0.3
599.5	Starches, inulin, gluten; albuminoidal substances; glues	10.9	16.8	–5.9
599.7	Other organic chemical products such as waxes, graphite, etc.	12.3	11.4	0.9

Notes: (i) Sources: Description of SITC items: See Ch.3, footnote 19.
Tariffs: See Appendix A.

(ii) In order to identify products by conviently short titles it sometimes proved necessary to use terms which are not part of the SITC description as set out in the table. In these cases, the shorthand description is added (in brackets and in quotation marks) after the full SITC title.

Chart 4.A.1 *UK and EEC Tariff Rates on Chemicals* (1965/67)

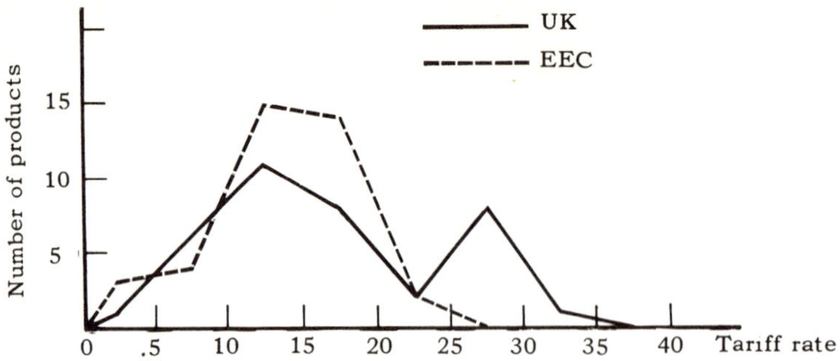

Source: Table 4.A.1.

Table 4.A.2 *Trade in Chemicals by Class (Mutual Trade Ranking)* (1965/67 average)

Class	No. of items	UK exports to EEC ($million) (1)	EEC exports to UK ($million) (2)	Intra-trade ($million) (3)	UK exports to third countries ($million) (4)	EEC exports to third countries ($million) (5)	(4)/(5) (per cent) (6)
I	9	62.7	24.4	356.4	237.0	434.9	54.5
II	4	32.8	23.6	251.1	53.9	140.1	38.5
III	6	35.5	34.6	271.5	135.7	341.9	39.7
IV	10	56.7	100.6	785.7	177.6	452.0	39.3
V	9	35.4	120.2	545.3	91.8	63.1	14.6
Total	38	223.0	303.4	2,210.0	696.1	1,999.4	34.8
Chemicals as % of all manufactures		10.9	16.3	11.6	10.7	14.6	–

Notes: See notes (i) to (v) of Table 4.A.3.

Chart 4.A.2 *Differences between Tariffs on Chemicals*
(UK minus EEC, 1965/67)

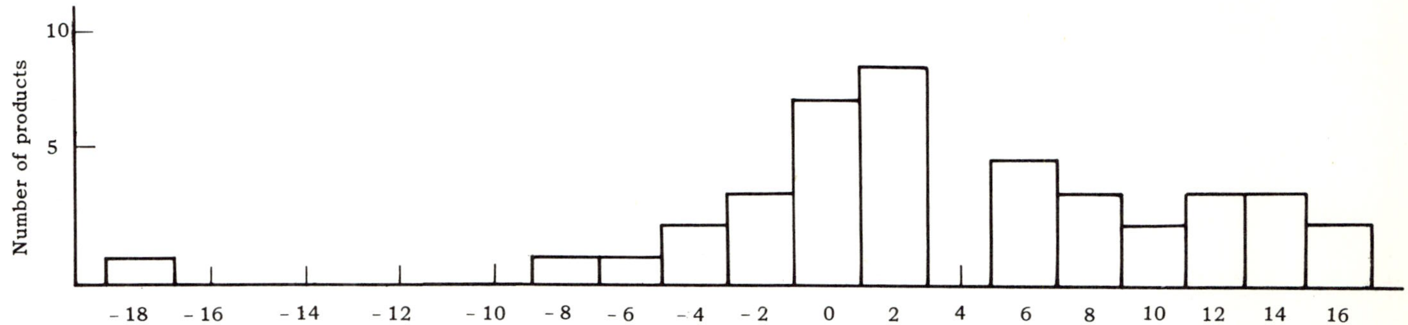

[a] The precise limits of the intervals are . . . –5 to –3.1, –3 to –1.1, –1 to +0.9, 1 to 2.9, 3 to 4.9, ...
Source: Table 4.A.1.

Table 4.A.3 *Trade in Chemicals: A Detailed Comparison*
(1965/67 average)

Class	SITC	Description	UK exports to EEC ($'000) (1)	EEC exports to UK ($'000) (2)	Intra-trade ($'000) (3)	UK exports to third countries ($'000) (4)	EEC exports to third countries ($'000) (5)	Special features (6)
I	533.1	Other colouring materials	6,072	2,660	23,890	17,447	14,976	
	533.2	Printing inks	983	385	7,321	4,743	4,379	
	533.3	Paints etc.	7,876	2,577	76,012	29,409	44,096	
	541.1	Vitamins etc.	3,489	835	21,083	1,527	10,791	V
	541.3	Antibiotics	11,067	3,334	35,063	9,243	22,536	III
	541.7	Medicaments	15,276	8,834	95,200	101,197	291,266	III
	554.1	Soaps	905	65	9,591	10,931	10,792	
	581.9	Other plastic materials	2,253	1,084	11,383	5,006	6,952	
	599.7	Other organic chemicals	14,750	4,652	76,880	57,505	29,096	
II	512.1	Hydrocarbons & derivatives	18,088	13,426	114,694	16,878	49,478	IV
	512.2	Alcohols	9,113	6,545	86,167	17,656	51,494	IV
	513.5	Metallic oxides	4,385	3,098	37,538	2,636	21,100	V
	571	Explosives	1,192	492	12,708	16,717	18,004	†
III	514.3	Other metallic salts (II)	3,953	4,502	28,120	10,094	25,470	*
	521	Mineral tar	3,012	2,939	21,332	6,636	7,260	I
	531.0	Synthetic dyes	11,744	11,880	82,233	40,258	111,673	
	553.0	Cosmetics	6,824	6,381	52,543	21,898	51,316	
	581.3	Cellulose & derivatives	6,364	5,590	47,542	26,142	35,840	
	599.2	Insecticides etc.	3,592	3,325	39,776	30,713	110,340	
IV	512.3	Ethers etc.	1,310	2,825	24,209	2,070	15,789	*
	512.4	Aldehydes etc.	1,114	2,765	26,279	5,572	16,719	
	513.2	Other chemical elements	3,233	8,478	39,870	3,148	26,996	*
	513.6	Other inorganic bases	3,047	4,826	41,303	18,896	49,489	
	514.2	Other metallic salts (I)	3,542	6,131	55,896	29,998	51,658	II
	532	Dyeing & tanning extracts	628	785	7,279	2,047	12,956	†
	554.2	Detergents	2,556	4,181	50,014	18,464	33,229	II
	561.2	Phosphatic fertilizers	361	2,054	36,930	186	14,827	
	581.1	Condensation products	18,741	28,071	182,869	43,619	74,579	II
	581.2	Polymers	22,125	40,468	321,030	53,638	155,762	

V	512.5	Organic acids	8,111	14,796	94,228	16,041	83,708	†
	512.7	Nitrogen compounds	8,233	25,995	108,692	15,692	159,753	
	512.8	Organo-inorganic compounds	10,538	19,688	91,690	15,112	91,242	
	513.3	Inorganic acids	796	3,801	26,414	1,765	10,573	
	514.9	Other inorganic chemicals	284	5,010	12,894	3,536	14,247	
	551	Essential oils	5,085	11,310	50,695	14,557	55,714	
	561.1	Nitrogenous fertilizers	2	10,476	30,410	15,647	124,451	
	561.3	Potassic fertilizers	4	12,573	64,297	23	61,436	
	599.5	Starches etc.	2,374	16,568	65,949	9,476	29,387	

Notes: (i) Sources: OECD *Commodity Trade*, Series C, and UN Statistical Papers, Series D, *Commodity Trade Statistics*.

(ii) Mutual trade and intra-trade (columns 1, 2 and 3) are valued c.i.f; exports to third countries (columns 4 and 5) are valued f.o.b.

(iii) All trade flows are shown as recorded, i.e., the tariff adjustment to mutual trade is not included.

(iv) The classing of commodities (shown with reference to mutual trade) is explained in Ch. 3, p. 31.

(v) For the definition of third countries see Ch. 2, p. 13.

(vi) The meaning of the entries in the 'Special features' column is as follows: a star indicates that the product 'just made' the class to which it is assigned, and a dagger that the product narrowly failed to reach the next higher class. The Roman numerals show in which class the products concerned – 'conflicting evidence' items – would be placed by the third country export rank order.

Table 4.A.4 *Chemicals: Determinants of Trade Diversion*

(a) In favour of UK sources of supply

Class	SITC	Description	EEC imports from third countries (1965/67 average; $'000)	Post-Kennedy CET (per cent)
I	533.1	Other colouring materials	3,611	9.9
	533.2	Printing inks	919	12.0
	533.3	Paints, etc.	11,214	10.8
	541.1	*Vitamins, etc.*	7,026	9.0
	541.3	*Antibiotics*	24,209	11.7
	541.7	*Medicaments*	15,037	17.0
	554.1	Soaps	298	12.0
	581.9	Other plastic materials	2,791	10.0
	599.7	Other organic chemicals	19,312	9.7
II	512.1	*Hydrocarbons & derivatives*	103,822	11.4
	512.2	*Alcohols*	40,630	10.3
	513.5	*Metallic oxides*	6,055	9.6
	571	Explosives	2,924	11.1
III	521	*Mineral tar*	15,621	2.5

(b) In favour of EEC sources of supply

Class	SITC	Description	UK imports from third countries (1965/67 average; $'000)	Post-Kennedy CET (per cent)
IV	512.3	Ethers, etc.	767	13.8
	512.4	Aldehydes, etc.	4,014	13.0
	513.2	Other chemical elements	9,911	5.3
	513.6	Other inorganic bases	11,202	10.1
	514.2	*Other metallic salts* (I)	2,025	9.4
	532	Dyeing & tanning extracts	3,553	7.0
	554.2	*Detergents*	3,476	12.0
	561.2	Phosphatic fertilisers	113	2.0
	581.1	*Condensation products*	24,564	12.8
	581.2	Polymers	32,298	15.5
V	512.5	Organic acids	20,561	13.1
	512.7	Nitrogen compounds	32,917	13.0
	512.8	Organo-inorganic compounds	21,722	13.1
	513.3	Inorganic acids	563	8.5
	514.9	Other inorganic chemicals	6,630	8.3
	551	Essential oils	12,590	7.5
	561.1	Nitrogenous fertilisers	215	8.0
	561.3	Potassic fertilisers	15,223	2.4
	599.5	Starches, etc.	20,849	10.8

Notes: (i) Sources: Third country imports: As for Table 4.A.3
Post-Kennedy Round Common External Tariffs: See Appendix A.
(ii) Descriptions printed in italics refer to conflicting evidence products.
(iii) Imports from third countries are valued c.i.f.
(iv) The classing of commodities (shown with reference to mutual trade) is explained in Chapter 3, p. 31.

particular group of products; the tariff rates themselves similarly do not appear to follow any regular pattern.

The overall trade picture is set out in Table 4.B.2. British exports to the EEC averaged $270 million, and EEC exports to the UK just over $200 million, so that the crude trade balance was in the UK's favour by a substantial margin. As usual both partners' exports to third countries were rather larger than mutual exports; the UK figure is $555 million, the EEC $990 million. The aggregate neutral market index is therefore 56.1. Taken overall, and before allowance for tariffs on mutual trade, the UK's performance for this industry group thus looks reasonably strong with respect both to mutual trade and to third country exports.

It will be noted that the total value of trade in class I products is very high in relation to that in the other classes. This is particularly true of the UK, not surprisingly, as the commodities concerned are those in which the UK is shown to be strongest, but it also holds in the case of the EEC's exports, especially those to the UK. As will be seen presently, the reason lies in the large size of trade in one particular item.

Brief mention must be made of column 6. It appears that the evidence supplied by the two performance tests is reasonably similar, except with regard to classes II and III, which the third country export ranking would place in inverse order.

The composition of the five performance classes is shown in Table 4.B.3. In view of the rather heterogenous nature of the goods being dealt with, examination of the table might best be conducted primarily with reference to particular commodity groups rather than to the individual classes. The first such group is leather (SITC 61). The picture which emerges is not very clear; three of the leather items (Cow Hide, Other Leather, and Fur Skins) are assigned to class I (though the last of these is very much a conflicting evidence commodity), whilst the other two (Calf Leather and Leather Manufactures) are found in classes V and IV respectively, and three other commodities which may perhaps be put alongside leather – Travel Goods, Leather Clothing, and Footwear[6] – are also assigned to the lower part of the table, the first to class IV and the other two to class V.

No special comment appears to be called for with regard to rubber goods (SITC 62 – two entries). In the case of wood and allied products (SITC 63) it is worth noting that whilst the semi-manufactured items (Veneer Sheets, and Plywood) are both found in class V, the more highly processed goods (Builders' Woodwork, and Furniture[7]) are assigned to class II, though the first of these is a conflicting evidence product and in any case quantitatively insignificant. Paper products (SITC 64) are scattered right through the list, and the existence of two conflicting evidence commodities (Newsprint, and Fibreboard), does little to clarify the picture. However, there appears to be some tendency for the quantitatively more important items to be assiged to classes II and III.

This takes the discussion to the final group of products, SITC 66 – mineral manufactures (other than metals). Once again there is no clear cut pattern except

6 SITC 831.0, 841.3 and 851.0 respectively.

7 SITC 821.0.

perhaps with regard to glass, as four out of the five products (Glass Containers, Glass Tableware, and the two semi-manufactures, Unworked Glass, and Ground or Polished Glass) are assigned to one or other of the last two classes, though it should be added that on the evidence of neutral market exports the last-mentioned commodity would join the fifth (Other Glass Articles) in class II.

One further product deserves comment – Diamonds, placed in class I. In most respects this is far and away the largest item in the list,[8] and it is thus not surprising that, as pointed out earlier, the aggregate value of class I is so much greater than that of the others. It is also clear that the result obtained for Diamonds must be interpreted with great caution, as it is not obvious in what sense the commodity can be said to be a product of UK manufacture.[9] In any event, the CET on Diamonds is very low; it thus appears that access to the EEC market is already quite easy and that UK entry would not greatly change the opportunities open to British exporters/traders.

The discussion is concluded with a brief look at the relevant third country imports and the post-Kennedy CET (Table 4.B.4) in order to assess the scope for trade diversion. Taking the upper part of the table (classes I and II) first one finds that in terms of size of EEC purchases from third countries Diamonds stand out as much the most promising product from the UK's point of view. However, the low level of the CET suggests that prospects may be rather more limited, quite apart from any other special factors which may apply. Substantial EEC purchases from neutral markets are also recorded for Other Leather, Other Printing Paper, and Other Paper, though in the case of the first of these the CET is once again fairly low.

Opportunities open to EEC producers for expanding exports as a result of trade diversion are indicated in the lower part of the table. Prospects look particularly good as regards Plywood, Kraft Paper, and Footwear, and possibly also Newsprint (a conflicting evidence product).[10]

C. Textiles and Clothing

The third industry section deals with textiles. The number of textile products seperately distinguished is 30; details of the composition of the industry, and the relevant UK and EEC tariffs are given in Table 4.C.1.

The tariff data are summarized in Charts 4.C.1 and 4.C.2. The first of these, which shows the frequency distributions of the 1965/67 tariff rates, indicates a relatively high level of tariff protection in both prospective partner areas (relative that is, to the overall UK and EEC positions respectively); the most common EEC

8 It is also the UK's most important export out of the 230 products.

9 Several of the points made about non-ferrous metals (pp. 67–68 below) probably also apply to Diamonds. There would thus be quite a strong case for excluding Diamonds from a list of manufactures to be subjected to the kind of analysis attempted in this study.

10 How far EEC accession on the part of the Scandinavian EFTA countries would affect the conclusions drawn in the last two paragraphs is a matter which, as explained earlier, cannot be resolved.

Table 4.B.1 *Manufactures of Leather, etc.: Products Covered and Tariff Rates*

SITC	Description	Tariffs (1965/67; per cent) UK	EEC	UK-EEC
611.3	Calf leather	20.0	9.0	11.0
611.4	Leather of other bovine cattle and equine leather ('Cow hide')	10.5	10.0	0.5
611.9	Leather of sheep, goats, etc. ('Other leather')	13.7	7.8	5.9
612	Manufactures of leather or of artificial or reconstituted leather	17.4	15.0	2.4
613.0	Fur skins, tanned or dressed including dyed	18.5	5.5	13.0
621.0	Materials of rubber	13.3	10.6	2.7
629.1	Rubber tyres and tubes for vehicles and aircraft	23.0	17.0	6.0
631.1	Veneer sheets	9.0	10.0	–1.0
631.2	Plywood including veneered panels	17.5	12.5	5.0
632.4	Builders' woodwork and prefabricated buildings of wood	15.0	14.0	1.0
641.1	Newsprint paper	15.8	7.8	8.0
641.2	Other printing and writing paper in rolls or sheets	14.0	14.0	–
641.3	Kraft paper and kraft paperboard	16.7	8.5	8.2
641.5	Machine-made paper and paperboard, simply finished, in rolls or sheets	16.7	16.0	0.7
641.6	Fibreboards and other building boards of wood pulp or of vegetable fibres	20.0	20.0	–
641.9	Other paper and paperboard in rolls or sheets	17.7	16.5	1.2
642.1	Paper bags, paperboard boxes and other containers of paper or paperboard	20.0	20.0	–
642.9	Articles of paper pulp, paper or paperboard, n.e.s.	17.0	17.0	–
661.2	Cement	7.5	8.0	–0.5
662	Clay construction materials and refractory construction materials ('Bricks and tiles')	18.5	10.5	8.0
663	Other mineral manufactures	15.0	10.9	4.1
664.3	Drawn or blown glass, unworked, in rectangles	15.0	10.0	5.0
664.4	Cast, rolled, drawn or blown glass in rectangles, surface ground or polished but not further worked	15.0	10.0	5.0
665.1	Carboys, bottles, jars, flasks and similar glass containers	26.3	21.0	5.3
665.2	Tableware and other household articles of glass	21.7	24.0	–2.3
665.8	Other articles made of glass	20.0	17.0	3.0
666.4	Porcelain or china household ware	15.0	10.0	5.0
667.2	Diamonds (other than industrial diamonds) not set or strung	10.0	3.0	7,0
821.0	Furniture	16.5	15.4	1.1
831.0	Travel goods, handbags, and similar articles	18.7	18.0	0.7
841.3	Apparel and clothing accessories of leather	27.5	15.7	11.8
851.0	Footwear	24.4	18.0	6.4

Notes: See Table 4.A.1.

Chart 4.B.1 *UK and EEC Tariffs on Manufactures of Leather, etc.* (1965/67)

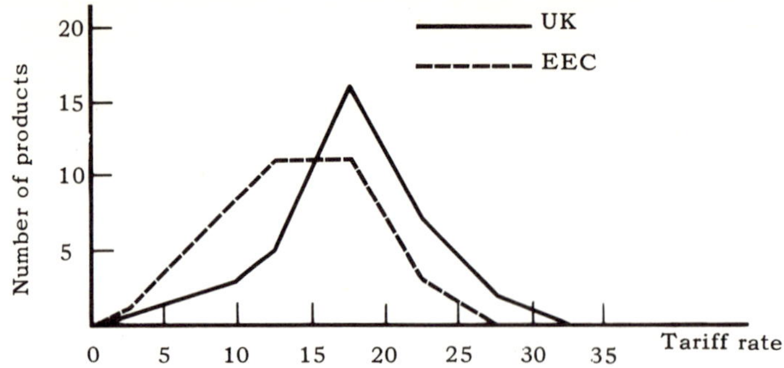

Source: Table 4.B.1.

Chart 4.B.2 *Differences between Tariffs on Manufactures of Leather, etc.* (UK minus EEC, 1965/67)

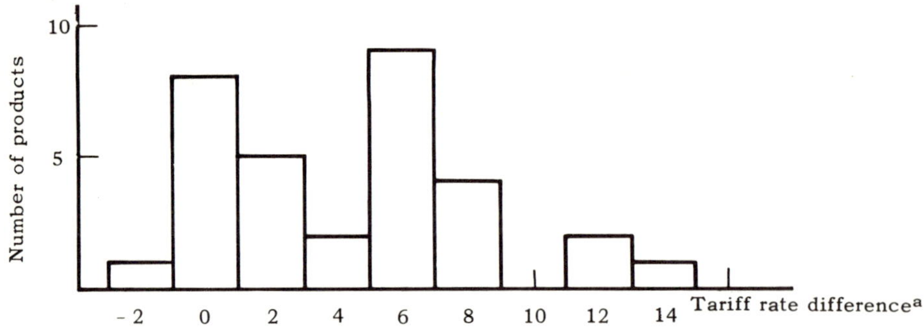

[a]See note (a) for Chart 4.A.2. Source: Table 4.B.1.

Table 4.B.2 *Trade in Manufactures of Leather,,etc. by Class (Mutual Trade Ranking)* (1965/67 *average*)

Class	No. of items	UK exports to EEC ($million) (1)	EEC exports to UK ($million) (2)	Intra-trade ($million) (3)	UK exports to third countries ($million) (4)	EEC exports to third countries ($million) (5)	(4)/(5) (per cent) (6)
I	7	206.4	101.0	626.2	331.4	296.7	111.8
II	6	34.9	22.9	509.3	73.0	178.6	40.9
III	5	16.9	17.6	290.0	68.3	151.7	45.0
IV	5	3.0	9.9	144.2	19.2	74.3	25.8
V	9	8.8	55.0	389.9	63.4	288.7	22.0
Total	32	269.8	206.4	1,959.6	555.2	990.0	56.1
Leather, etc. as % of all manufactures		13.2	11.1	10.3	8.5	7.2	–

Notes: See notes (i) to (v) of Table 4.A.3.

(1965/67 average)

Class	SITC	Description	UK exports to EEC ($'000) (1)	EEC exports to UK ($'000) (2)	Intra-trade ($'000) (3)	UK exports to third countries ($'000) (4)	EEC exports to third countries ($'000) (5)	Special features (6)
I	611.4	Cow hide	3,558	432	38,498	8,863	11,543	
	611.9	Other leather	13,468	1,553	66,368	12,905	13,449	
	613.0	Fur skins	8,371	2,222	32,479	2,297	7,305	IV
	621.0	Rubber materials	7,935	4,631	50,479	14,949	25,554	*
	642.9	Other paper manufactures	6,393	2,493	57,923	21,328	29,820	
	663	Other mineral manufactures	17,563	6,871	113,971	22,521	32,217	
	667.2	Diamonds	149,089	827,772	266,451	248,529	176,802	
II	632.4	Builders' woodwork	464	79	17,061	516	3,516	V
	641.2	Other printing paper	2,295	1,362	77,877	17,357	14,845	*
	641.9	Other paper	13,688	8,770	92,684	13,197	29,335	
	662	Bricks & tiles	5,400	5,144	108,711	12,590	69,877	*V
	665.8	Other glass articles	2,285	1,728	15,534	6,163	11,862	
	821.0	Furniture	10,722	5,853	197,464	23,165	49,176	
III	629.1	Rubber tyres	10,102	8,525	151,009	37,894	107,500	
	641.1	Newsprint	19	112	12,021	151	2,785	V
	641.5	Machine-made paper	4,359	5,239	60,713	10,071	17,426	
	642.1	Paper bags etc.	1,101	973	36,256	5,170	15,605	
	666.4	Porcelain	1,293	2,754	29,989	15,056	8,421	I
IV	612	Leather manufactures	768	1,281	19,375	4,253	10,072	
	641.3	Kraft paper	133	524	18,281	563	3,207	†
	661.2	Cement	237	2,667	37,539	5,489	13,998	
	665.1	Glass containers	484	1,683	30,297	3,833	17,312	
	831.0	Travel goods	1,331	3,761	38,754	5,030	26,669	
V	611.3	Calf leather	1,140	2,790	2,093	3,218	21,276	
	631.1	Veneer sheets	1,001	5,279	37,332	837	8,863	
	631.2	Plywood	136	9,267	27,305	287	6,354	
	641.6	Fibreboard	67	1,171	14,443	523	1,172	II
	664.3	Unworked glass	979	3,888	19,224	3,546	31,503	
	664.4	Ground or polished glass	1,059	2,754	21,851	11,037	16,993	II
	665.2	Glass tableware	1,205	5,324	46,120	4,763	27,155	
	841.3	Leather clothing	171	1,156	12,247	2,135	24,973	
	851.0	Footwear	3,032	23,383	209,252	37,004	150,422	

Notes: See notes to Table 4.A.3.

Table 4.B.4 *Manufactures of Leather etc.: Determinants of Trade Diversion*

(a) In favour of UK sources of supply

Class	SITC	Description	EEC imports from third countries (1965/67 average; $'000)	Post-Kennedy CET (per cent)
I	611.4	Cow hide	7,366	8.0
	611.9	Other leather	49,806	3.9
	613.0	*Fur skins*	12,154	2.5
	621.0	Rubber materials	8,863	6.0
	642.9	Other paper manufactures	12,579	10.6
	663	Other mineral manufactures	19,603	6.4
	667.2	Diamonds	154,523	3.7
II	632.4	*Builders' woodwork*	3,732	6.3
	641.2	Other printing paper	37,018	9.5
	641.9	Other paper	23,065	12.6
	662	*Bricks & tiles*	15,854	5.7
	665.8	Other glass articles	4,493	8.5
	821.0	Furniture	11,149	8.1
III	666.4	*Porcelain*	8,783	5.0

(b) In favour of EEC sources of supply

Class	SITC	Description	UK imports from third countries (1965/67 average; $'000)	Post-Kennedy CET (per cent)
III	641.1	*Newsprint*	77,196	7.0
IV	612	Leather manufactures	584	7.7
	641.3	Kraft paper	71,948	7.5
	661.2	Cement	808	4.0
	665.1	Glass containers	425	10.5
	831.0	Travel goods	5,756	11.3
V	611.3	Calf leather	3,602	0
	631.1	Veneer sheets	956	3.5
	631.2	Plywood	114,899	9.0
	641.6	*Fibreboard*	10,029	11.0
	664.3	Unworked glass	466	6.0
	664.4	*Ground or polished glass*	133	5.0
	665.2	Glass tableware	4,323	15.5
	841.3	Leather clothing	1,607	8.7
	851.0	Footwear	26,947	12.0

Notes: See notes to Table 4.A.4.

tariff falls in the range 15 per cent to 19.9 per cent, whilst the UK distribution has its (rather less marked) peak in the next range, 20 per cent to 24.9 per cent. Table 4.C.1 shows that there is some tendency for both UK and EEC tariff rates to rise as one moves down the list from the comparatively basic semi-manufactures like the yarns to more finished items, such as clothing. The distribution of tariff differences (Chart 4.C.2) is pretty even over most of the scale, with a slight peak in the +4 range. British tariffs which exceeded their EEC equivalents by relatively large margins (5 percentage points or more) include the four clothing items as well as the three rates referring to SITC 655 (Specialized Textile Products), whilst in the case of the yarns and the woven fabrics UK and EEC tariffs were on the whole reasonably similar (for two-thirds of the products concerned the tariff differences lie in the –2 to +2 range).

Table 4.C.2 summarizes mutual and third country trade in textiles. UK exports to the EEC on average came to $145 million, and EEC exports to the UK to just over $20 million more; in other words, the UK's crude trade balance was negative. Exports to third countries amounted to $405 million for the UK and $1,033 million for the EEC; the neutral market index works out at just under 40. The aggregate performance of the UK in this group of products was thus not very good on either of the two criteria adopted.

Column 6 suggests that there is only partial agreement between the two performance measures. Relative third country exports of the products in the first three classes follow the mutual trade ranking, but then the third country export ratio begins to rise again instead of declining further. It is thus not surprising that Table 4.C.3 (which sets out the trade performances in detail) shows three out of the five items placed in class V to be conflicting evidence products, with two of these (Bleached Cotton Yarn, and Canvas Goods) performing so well in third country trade that the alternative rank order assigns them to class I. The third conflicting evidence product in class V is another cotton product, namely Bleached Cotton Fabrics, though in this case the neutral market ranking makes a difference of only two classes.

The composition of class IV presents a rather more clear cut picture, with only one conflicting evidence product (Other Textile Articles). Particularly noteworthy is perhaps the uncontested assignment to that class of both Synthetic Yarns and Synthetic Fabrics. With the other artificial fabric item (Regenerated Fibre Fabrics) placed in class V and doubts about the position of the corresponding yarn entry (Regenerated Fibre Yarn, assigned to class I by the mutual trade ranking but to class III by the third country export record), the indications are that British producers of artificial fibres may occupy a relatively weak position in an enlarged EEC.

Class III is rather a large group, with nearly one-third of all items assigned to it. The majority of the commodities concerned are finished products. Two of these are conflicting evidence items, the first, Other Carpets, being placed by the third country rank order in class I, the second, Knitwear, in class V. Most of the products in class II are also finished goods (including one conflicting evidence item) and quantitatively rather unimportant.

Class I consists of only four entries, including one conflicting evidence case (Regenerated Fibre Yarn). Special attention may be drawn to Woollen Fabrics (quantitatively the third most important product in the list), in view of the UK's particularly strong third country export performance; UK exports to neutral markets were almost twice those of the EEC, as compared to a ratio for all class I products (neutral market export ranking) of just under 100. In this connection it may also be noted that Wool Yarn, the second wool item, is placed at the upper end of class II and thus enjoys quite a strong position; on the other hand, the neutral market export record of Wool Yarn is rather poor.

The data used for examining the scope for trade diversion is assembled in Table 4.C.4; the upper part of the table – showing EEC third country imports of goods placed in classes I and II, together with one class III item – is taken first. The initial impression is that two products stand out as providing good opportunities for British exporters – Knotted Carpets and Bags and Sacks, in that order. However, the second of these is very much a conflicting evidence product, so that it is not clear that British producers would have the necessary edge over their EEC rivals, and the first is a special case, in the sense that the description 'Knotted Carpets' appears largely (perhaps wholly) to refer to oriental rugs, i.e., to a commodity which is not in fact produced either in the UK or in the EEC.[11] In the circumstances there is clearly little sense in speaking of British firms facing good opportunities for raising their exports of Knotted Carpets to the EEC by way of underselling producers in third countries (or, for that matter, by means of replacing EEC 'producers' in their own market).

The scope for EEC producers benefiting from trade diversion seems quite good. The lower half of the table – showing UK third country imports of the items in classes IV and V, and also one class III product – contains three goods (Grey Cotton Fabrics, Bleached Cotton Fabrics, and Knitwear) which, on the basis of past evidence, the UK buys in substantial amounts from third countries and which the CET subjects to a relatively high tariff rate. However, once again the case is not straightforward. (a) The last two of the goods mentioned are conflicting evidence products, so that EEC producers may not be in a position fully to exploit the opportunities which in principle exist. (b) With regard to Grey as well as Bleached Cotton Fabrics the UK's main sources of supply among third countries are India, Pakistan, and Hong Kong, and it is well established that these countries are pressing for arrangements to be made which would keep the British market open to their textile products. In so far as the entry agreement between the UK and the EEC includes special provisions to safeguard the trade interests of the countries mentioned, trade diversion opportunities open to EEC producers will diminish.[12]

D. Iron and Steel

The fourth industry, the iron and steel sector, is for present purposes distinguished by two special features. In the first place the commodities included form a comparatively homogeneous group, and many of them could be produced in one and the same plant; *a priori* one might therefore expect comparative cost differences between them to be relatively limited in extent. Secondly, as far as producers in the Six are concerned, many steel products do not, strictly speaking, fall within

11 This hypothesis is suggested by the fact that both the EEC's and the UK's imports from third country sources of supply primarily originate from countries which produce oriental carpets – Iran, India, Pakistan, and Afghanistan – and derives indirect support from the low level of mutual trade and the miniscule size of exports to neutral markets (Table 4.C.E).

12 In general it should be said that with regard to cotton textiles in particular world trade arrangements may alter in ways which will seriously upset the conclusions drawn from the data.

Table 4.C.1 *Textiles and Clothing: Products Covered and Tariff Rates*

SITC	Description	Tariffs (1965/67; per cent) UK	EEC	UK-EEC
651.2	Yarn of wool and animal hair	12.0	7.6	4.4
651.3	Cotton yarn and thread, grey, not mercerized nor put up for retail sale	16.0	9.0	7.0
651.4	Cotton yarn and thread, bleached, dyed, mercerized, etc.	10.0	12.5	–2.5
651.5	Yarn and thread of flax, ramie and true hemp	12.0	12.0	–
651.6	Yarn and thread of synthetic fibres	16.0	15.7	0.3
651.7	Yarn and thread of regenerated (artificial) fibres	16.0	14.0	2.0
652.1	Cotton fabrics, woven, grey, not mercerized	20.0	17.0	3.0
652.2	Cotton fabrics, woven, other than grey	20.0	17.5	2.5
653.1	Silk fabrics, woven	20.8	17.4	3.4
653.2	Woollen fabrics, woven (including fabrics of fine hair)	17.5	17.0	0.5
653.3	Linen, ramie and true hemp fabrics, woven	18.8	20.5	–1.7
653.4	Jute fabrics woven	20.0	23.0	–3.0
653.5	Fabrics, woven, of synthetic fibres	17.5	17.0	0.5
653.6	Fabrics, woven, of regenerated (artificial) fibres	17.5	19.5	–2.0
653.7	Knitted or crocheted fabrics, not elastic nor rubberized	21.3	17.0	4.3
654.0	Tulle, lace, embroidery, ribbons, trimmings and other small wares	22.0	17.7	4.3
655.4	Coated or impregnated textile fabrics and products, n.e.s.	20.8	14.8	6.0
655.6	Cordage, cables, ropes, twines and manufactures thereof	24.0	16.0	8.0
655.8	Wadding, wicks and textile fabrics for use in machinery or plant	20.0	11.0	9.0
656.1	Bags and sacks of textile materials	25.0	19.0	6.0
656.2	Tarpaulins, tents, awnings, sails, other made-up canvas goods	21.0	19.0	2.0
656.6	Blankets, travelling rugs and coverlets	26.8	19.0	7.8
656.9	Other made-up articles of textile materials	26.0	22.5	3.5
657.4	Linoleum and similar floor coverings	16.7	17.5	–0.8
657.5	Carpets, carpeting and rugs, knotted	33.0	32.0	1.0
657.6	Other carpets, carpeting and rugs	25.0	22.0	3.0
841.1	Clothing of textile fabric, not knitted or crocheted ('Clothing excluding knitwear')	26.9	18.2	8.7
841.2	Clothing accessories of textile fabric, not knitted or crocheted	25.8	20.3	5.5
841.4	Clothing and accessories, knitted or crocheted ('Knitwear')	25.1	20.0	5.1
841.5	Headgear	24.7	16.5	8.2

Notes: See Table 4.A.1.

Chart 4.C.1 *UK and EEC Tariff Rates on Textiles and Clothing* (1965/67)

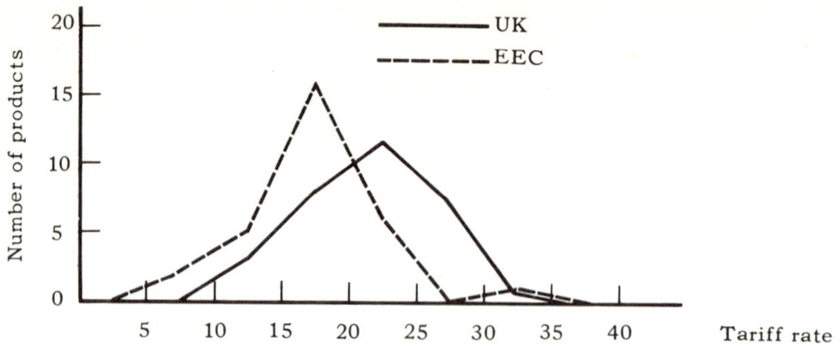

Source: Table 4.C.1.

Chart 4.C.2 *Differences between Tariffs on Textiles and Clothing* (UK minus EEC, 1965/67)

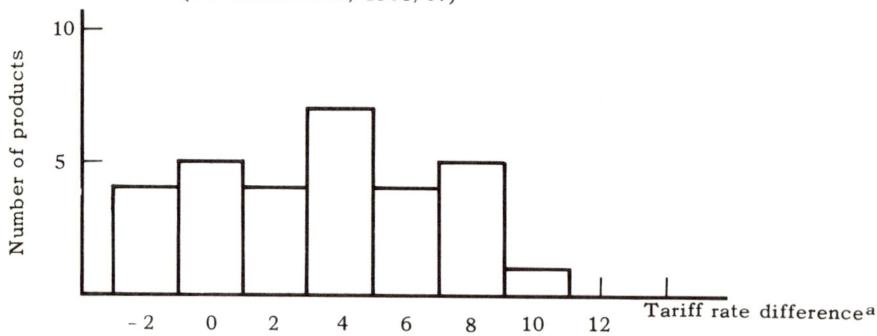

[a]See note (a) for Chart 4.A.2. Source: Table 4.C.1.

Table 4.C.2 *Trade in Textiles and Clothing by Class (Mutual Trade Ranking)* (1965/67 average)

Class	No. of items	UK exports to EEC ($million) (1)	EEC exports to UK ($million) (2)	Intra-trade ($million) (3)	UK exports to third countries ($million) (4)	EEC exports to third countries ($million) (5)	(4)/(5) (per cent) (6)
I	4	53.8	21.3	336.0	112.6	92.8	121.4
II	7	24.8	8.1	315.0	49.8	124.8	39.9
III	9	44.9	41.5	1,052.4	107.6	386.9	27.8
IV	5	13.8	37.4	485.6	60.0	212.1	28.3
V	5	7.9	58.5	335.4	74.5	216.5	34.4
Total	30	145.3	166.9	2,524.5	404.6	1,033.2	39.2
Textiles as % of all manufactures		7.1	8.9	13.2	6.2	7.5	–

Notes: See notes (i) to (v) of Table 4.A.3.

(1965/67 average)

Class	SITC	Description	UK exports to EEC ($'000) (1)	EEC exports to UK ($'000) (2)	Intra-trade ($'000) (3)	UK exports to third countries ($'000) (4)	EEC exports to third countries ($'000) (5)	Special features (6)
I	651.7	Regenerated fibre yarn	7,447	2,497	69,328	13,952	40,363	III
	653.2	Woollen fabrics	41,249	17,953	255,821	84,323	43,279	
	653.3	Linen fabrics	1,365	306	5,201	14,097	8,704	
	657.5	Knotted carpets	3,703	543	5,684	271	463	
II	651.2	Wool yarn	12,313	1,693	172,066	16,741	49,172	† IV
	651.5	Flax yarn	2,422	1,516	16,176	3,810	2,579	
	655.4	Coated fabrics	5,701	2,455	61,376	7,937	21,482	†
	655.6	Cord, cables, etc.	482	263	12,747	8,120	16,271	
	655.8	Wadding, etc.	2,237	1,577	15,534	6,163	11,862	
	656.1	Bags & sacks	435	149	11,450	1,678	13,884	V
	657.4	Linoleum	1,252	493	25,646	5,333	9,594	
III	651.3	Grey cotton yarn	590	1,102	42,589	4,382	9,754	
	653.4	Jute fabrics	239	227	13,423	1,979	4,163	
	653.7	Knitted fabrics	2,522	2,485	118,291	8,879	35,433	
	654.0	Small wares	2,125	2,635	50,298	9,394	26,475	
	656.6	Blankets	753	705	15,463	2,813	11,704	
	657.6	Other carpets	6,166	4,733	122,696	14,859	18,653	† I
	841.1	Clothing exc. knitwear	15,141	13,200	303,821	29,954	74,465	
	841.4	Knitwear	15,325	14,232	365,987	30,359	193,787	V
	841.5	Headgear	2,077	2,216	19,858	5,024	12,427	
IV	651.6	Synthetic yarn	8,338	19,099	229,475	27,892	121,253	
	652.1	Grey cotton fabrics	696	2,080	36,359	2,150	4,941	
	653.5	Synthetic fabrics	1,277	8,895	130,285	14,998	48,151	
	656.9	Other textile articles	1,274	3,821	39,465	9,932	19,976	* II
	841.2	Clothing accessories	2,244	3,551	50,000	5,064	17,824	
V	651.4	Bleached cotton yarn	220	1,588	16,550	14,488	15,864	† I
	652.2	Bleached cotton fabrics	3,186	26,697	131,901	43,878	113,565	III
	653.1	Silk fabrics	477	1,951	16,494	1,563	15,166	
	653.6	Regen. fibre fabrics	3,825	26,426	162,026	12,001	69,011	
	656.2	Canvas goods	232	1,816	8,449	2,536	2,929	I

Notes: See notes to Table 4.A.3.

Table 4.C.4 *Textiles & Clothing: Determinants of Trade Diversion*

(a) In favour of UK sources of supply

Class	SITC	Description	EEC imports from third countries (1965/67 average; $'000)	Post-Kennedy CET (per cent)
I	651.7	*Regenerated fibre yarn*	4,395	8.9
	653.2	Woollen fabrics	2,875	15.3
	653.3	Linen fabrics	1,071	13.3
	657.5	Knotted carpets	69,869	19.0
II	651.2	*Wool yarn*	827	6.7
	651.5	Flax yarn	635	8.1
	655.4	Coated fabrics	10,509	10.4
	655.6	Cord, cables, etc.	2,503	11.3
	655.8	Wadding, etc.	4,493	6.0
	656.1	*Bags & Sacks*	19,878	12.0
	657.4	Linoleum	630	11.0
III	657.6	*Other carpets*	9,806	16.0

(b) In favour of EEC sources of supply

Class	SITC	Description	UK imports from third countries (1965/67 average; $'000)	Post-Kennedy CET (per cent)
III	841.4	*Knitwear*	41,573	15.4
IV	651.6	Synthetic yarn	11,548	10.6
	652.1	Grey cotton fabrics	69,010	12.5
	653.5	Synthetic fabrics	5,351	14.5
	656.9	*Other textile articles*	*18,301*	14.5
	841.2	Clothing accessories	10,147	12.0
V	651.4	*Bleached cotton yarn*	154	9.0
	652.2	*Bleached cotton fabrics*	23,423	13.4
	653.1	Silk fabrics	2,111	11.5
	653.6	Regenerated fibre fabrics	8,698	15.3
	656.2	*Canvas goods*	885	15.5

Notes: See notes to Table 4.A.4.

the scope of the EEC, but rather that of its precursor, the European Coal and Steel Community (ECSC). However, there can be little doubt that the UK's accession to the EEC would be accompanied by her also gaining entry to the ECSC, and apart from a point relating to tariffs – taken up immediately below – the difference can be disregarded. Subsequent discussion will therefore refer to ECSC products as if they fell within the scope of the EEC.

It may be noted in passing that the absence of internal trade barriers on many of the steel products since the early 1950s should in principle at least mean that the process of specialization among the Six has been carried much further for these commodities than for the remaining manufactures produced by the EEC countries. *Ipso facto* the pattern of trade between the EEC countries and outsiders (including the UK) should be less distorted by the process of transition to free trade conditions within the Common Market than may be the case for EEC products generally (see p. 20 above). It does not, however, follow that the results for the steel industry (except perhaps those referring to the position of steel products relative to one another) can be regarded as particularly reliable, because the whole exercise is a comparative one, and distortions affecting non-steel goods are bound also to reduce the accuracy of the evidence relating to the position of this sector.

The 24 products covered, together with the relevant UK and EEC tariff rates for 1965/67, are set out in Table 4.D.1. Before the tariff data are discussed it should be pointed out that a number of EEC duties are not part of the CET, for unlike the Rome Treaty the instrument setting up the ECSC (the Treaty of Paris) did not provide for a common tariff vis-à-vis third countries. However, following a recommendation by the then High Authority of the ECSC the six member countries in 1964 aligned the great majority of their steel tariffs to a uniform set of rates (generally that prescribed in the Italian customs schedule), so that over the 1965/67 period the EEC countries in fact largely imposed common rates of duty on steel imports.[13]

The frequency distributions of UK and EEC steel tariffs are set out in Chart 4.D.1. Two points stand out – (a) the EEC's tariffs on the 24 products were rather less divergent than the UK's, and (b) both in Britain and in the EEC customs duties on imports of steel were rather lower than their respective tariff barriers against imports of manufactures generally. The figures in Table 4.D.1 indicate that as in the case of textiles, UK as well as EEC tariff rates tended to be rather lower for the basic products in the upper part of the list than they were for the more highly processed items further down.

Nonetheless, as Chart 4.D.2 shows, differences between UK and EEC steel duties extend over a wide range – from –8 to +8. The tariff differences form a surprisingly regular pattern. Commodities for which EEC tariffs exceeded UK rates of duty are almost without exception basic steel products; *mutatis mutandis*, cases of British tariffs exceeding EEC rates virtually all refer to rather more finished steel items, such as wire and tubes.

Data showing UK and EEC trade performances in the aggregate and by class are set out in Table 4.D.2. Mutual UK/EEC trade in steel was quite low both in absolute terms (UK exports to the EEC averaged $65 million, EEC exports to the UK $88 million) and as a proportion of total trade in manufactures between the prospective partners (as shown in the last row). Neutral market exports were rather more sizeable; the UK figure is $340 million, that for the EEC just under $1,250 million. The overall neutral market export ratio works out at 27.3, much the lowest figure for any of the nine industry groups. As the crude trade balance was also in deficit the aggregate figures convey rather a pessimistic impression from the UK point of view.

It can be seen from column 6 that the neutral market export ratio for class IV is significantly higher even than that for class II, suggesting that there is only a limited measure of agreement between the two performance indicators. This impression is confirmed by the detailed trade performance data set out in

13 Any remaining differences were averaged, by finding the unweighted arithmetic mean of national rates.

Table 4.D.3. Exactly half the 24 items are conflicting evidence products, much the highest proportion for any of the nine industry groups.

The incidence of the conflicting evidence cases is wholly confined to classes II and III (which are, of course, much the largest classes). As many as ten products move at least two classes down on the evidence of third country trade, and only two products move up to class I. It thus appears that in terms of third country exports the steel industry's performance was considerably worse than it was with reference to mutual trade.

Reference has just been made to the exceedingly uneven distribution of the 24 items over the five classes, with only five products assigned to classes I, IV and V taken together, and the remainder in classes II and III. This similarity in the performance of most of the steel items may at first sight be interpreted as confirmation of a point made at the beginning of the section – that the goods under consideration are reasonably homogeneous and that one might therefore expect comparative cost differences between them to be rather small. However, it must be admitted that the presence of so many conflicting evidence cases in classes II and III rather places this interpretation on its head.[14]

Table 4.D.3 also shows that there are marked differences between the items in the size of mutual trade. In particular, one single product (Thin Uncoated Plates) accounts for more than one-fifth of UK exports to the EEC and nearly two-fifths of EEC exports to the UK. It is true that in terms of the usual indicator of the importance of different commodities – the size of intra-trade – Thin Uncoated Plates are also the largest item, but the variations in the size of intra-trade are much less pronounced. It is not obvious what meaning should be attached to this aspect of the results.

In view of the discrepancies in the evidence for so many items there appears to be little merit in going through Table 4.D.3 in any further detail. By the same token no useful purpose would be served by any lengthy discussion of opportunities for trade diversion, at least as far as UK producers are concerned, for as Table 4.D.4 shows, the two products which the EEC imported in large quantities from third countries are both conflicting evidence cases. For EEC producers the outlook regarding gains from trade diversion is only a little less obscure; there is just one item (Other Ferro-Alloys) which the UK imported in significant amounts from third countries and whose assignment to class V is not contested. On the other hand, the level of the CET on this product is not particularly high, even by comparison with other steel tariffs.

E. *Non-Ferrous Metals and Miscellaneous Metal Manufactures*

The 24 items brought together in the fifth industry section can be divided into two distinct groups. There are 14 products assigned to SITC 68 (Non-Ferrous Metals), whilst the remaining 10 belong to SITC 69 (Manufactures of Metal not elsewhere specified); the latter thus include articles made of iron and steel

14 A list of steel products by class drawn up on the basis of neutral market performance ranking indicates a somewhat more even distribution of the 24 items, with 3 in class I, 2 in class II, 4 in class III, 7 in class IV, and 8 in class V.

Table 4.D.1 *Iron and Steel: Products Covered and Tariff Rates*

SITC	Description	Tariffs (1965/67; per cent) UK	EEC	UK-EEC
671.2	Pig iron (including cast iron)	0	5.0	−5.0
671.4	Ferro-manganese	0	6.5	−6.5
671.5	Other ferro-alloys	0	7.3	−7.3
672.3	Ingots of iron or steel	5.0	9.5	−4.5
672.5	Blooms, billets, slabs, sheet bars and roughly forged pieces of iron and steel	12.7	8.4	4.3
672.7	Iron or steel coils for re-rolling	5.0	10.0	−5.0
673.1	Wire rod of iron or steel	7.3	9.5	−2.2
673.2	Bars and rods of iron or steel; hollow mining drill steel	7.3	9.7	−2.4
673.4	Angles, shapes and sections, 80 mm or more, and sheet piling of iron or steel	12.3	9.4	2.9
673.5	Angles, shapes and sections, less than 80 mm, of iron or steel	11.0	9.5	1.5
674.1	Universals and heavy plates and sheets, more than 4.75 mm in thickness	10.5	9.5	1.0
674.2	Medium plates and sheets, 3 mm to 4.75 mm in thickness	11.0	9.7	1.3
674.3	Plates and sheets, less than 3 mm in thickness, uncoated ('Thin uncoated plates')	11.0	9.5	1.5
674.7	Tinned plates and sheets	10.0	9.6	0.4
674.8	Plates and sheets, less than 3 mm in thickness, coated (excluding tinned plates or sheets) ('Thin coated plates')	11.0	9.5	1.5
675.0	Hoop and strip of iron or steel	9.0	9.7	−0.7
676	Rails and railway track construction material of iron or steel	12.5	18.0	−5.5
677.0	Iron and steel wire (excluding wire rod)	17.0	9.7	7.3
678.1	Tubes and pipes of cast iron	17.5	13.0	4.5
678.2	Tubes and pipes of iron or steel, seamless	20.0	14.0	6.0
678.3	Tubes and pipes of iron or steel, welded, clinched, etc.	20.0	14.0	6.0
678.4	High pressure hydro-electric conduits of steel	20.0	13.0	7.0
678.5	Tube and pipe fittings of iron or steel	17.5	14.0	3.5
679	Iron and steel castings and forgings, unworked, n.e.s.	20.0	13.0	7.0

Notes: See Table 4.A.1.

Chart 4.D.1 *UK and EEC Tariff Rates on Iron and Steel* (1965/67)

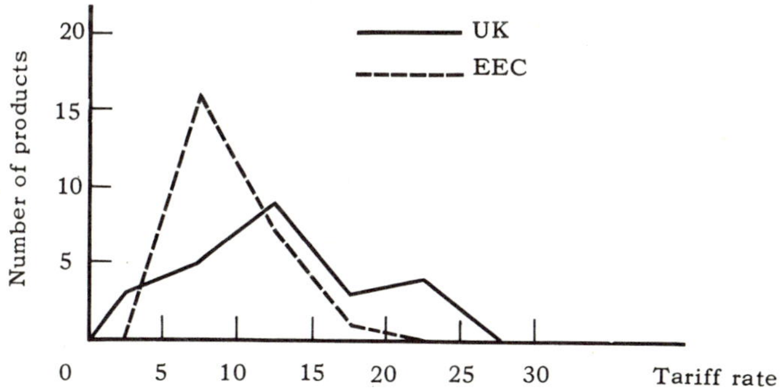

Source: Table 4.D.1.

Chart 4.D.2 *Differences between Tariffs on Iron and Steel* (UK minus EEC, 1965/67)

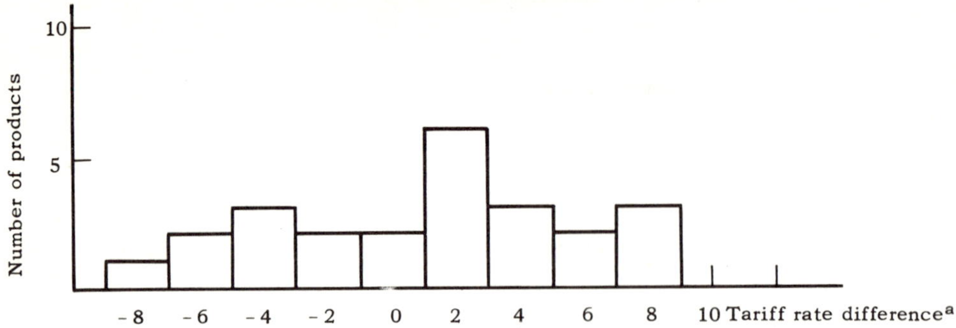

[a]See note (a) for Chart 4.A.2.   Source: Table 4.D.1.

Table 4.D.2 *Trade in Iron and Steel by Class (Mutual Trade Ranking)* (1965/67 average)

Class	No. of items	UK exports to EEC ($million) (1)	EEC exports to UK ($million) (2)	Intra-trade ($million) (3)	UK exports to third countries ($million) (4)	EEC exports to third countries ($million) (5)	(4)/(5) (per cent) (6)
I	1	1.4	1.0	12.6	10.3	15.5	66.5
II	10	33.5	11.7	706.5	145.5	527.7	27.6
III	9	14.4	21.9	763.1	123.9	494.5	25.1
IV	1	13.6	33.8	364.3	56.0	177.2	31.6
V	3	2.4	21.0	84.1	3.9	31.2	12.5
Total	24	65.3	88.5	1,930.6	339.6	1,246.0	27.3
Iron and steel as % of all manufactures		3.2	4.7	10.1	5.2	9.1	–

Notes: See notes (i) to (v) of Table 4.A.3.

Table 4.D.3 *Trade in Iron and Steel: A Detailed Comparison*

(1965/67 average)

Class	SITC	Description	UK exports to EEC ($'000) (1)	EEC exports to UK ($'000) (2)	Intra-trade ($'000) (3)	UK exports to third countries ($'000) (4)	EEC exports to third countries ($'000) (5)	Special features (6)
I	676	Rails	1,412	95	12,622	10,286	15,457	
II	671.2	Pig iron	1,566	398	35,255	1,701	15,111	V
	672.5	Blooms & billets	2,250	1,059	70,325	7,353	24,217	* IV
	672.7	Coils	2,079	5	105,448	3,139	39,499	V
	673.5	Angles etc. less than 80 mm	3,334	855	74,550	3,924	89,253	V
	674.1	Universals	6,176	586	151,079	22,219	74,159	IV
	674.2	Medium plates	1,494	383	43,704	5,592	13,894	
	674.7	Tinplate	2,131	1	58,223	39,301	50,570	
	677.0	Wire	1,728	631	47,576	18,694	68,455	IV
	678.2	Seamless tubes	8,863	5,528	81,059	32,148	127,624	IV
	678.5	Tube pipings	3,834	2,263	39,269	11,421	24,915	
III	672.3	Ingots	148	28	20,061	516	267	† I
	673.1	Wire rod	1,950	4,370	113,116	6,222	54,820	* V
	673.2	Bars and rods	3,952	7,721	253,558	30,520	160,386	V
	673.4	Angles etc. 80 mm or more	1,015	1,269	118,310	24,401	60,796	
	674.8	Thin coated plates	1,698	1,511	45,787	20,530	59,138	
	675.0	Hoop & strip	1,803	4,015	123,831	16,084	40,662	
	678.1	Cast iron tubes	127	89	5,538	5,548	16,020	
	678.3	Welded tubes	3,323	2,415	75,018	16,664	98,612	† V
	679	Castings	413	506	7,840	3,439	3,808	* I
IV	674.3	Thin uncoated plates	13,620	33,773	364,341	56,012	177,179	
V	671.4	Ferro-manganese	690	3,762	27,046	1,309	12,880	
	671.5	Other ferro-alloys	1,658	7,232	28,864	2,545	11,442	
	678.4	Hydro-electric conduits	65	9,967	28,224	38	6,862	

Notes: See notes to Table 4.A.3.

Table 4.D.4 *Iron and Steel: Determinants of Trade Diversion*

(a) In favour of UK sources of supply

Class	SITC	Description	EEC imports from third countries (1965/67 average; $'000)	Post-Kennedy CET (per cent)
I	676	Rails	182	6.0
II	671.2	*Pig iron*	29,796	4.0
	672.5	*Blooms and billets*	4,621	6.0
	672.7	*Coils*	46,298	7.2
	673.5	*Angles etc. less than 80 mm*	994	7.0
	674.1	*Universals*	7,775	6.7
	674.2	Medium plates	1,985	6.7
	674.7	Tinplate	4,633	6.7
	677.0	*Wire*	1,487	7.3
	678.2	*Seamless tubes*	7,658	9.0
	678.5	Tube fittings	10,296	10.0
III	672.3	*Ingots*	3,601	6.0
	679	*Castings*	410	7.0

(b) In favour of EEC sources of supply

Class	SITC	Description	UK imports from third countries (1965/67 average; $'000)	Post-Kennedy CET (per cent)
III	673.1	*Wire rod*	958	6.7
	673.2	*Bars & rods*	2,895	7.3
	678.3	*Welded tubes*	2,658	10.0
IV	674.3	Thin uncoated plates	8,729	6.9
V	671.4	Ferro-manganese	6,566	6.0
	671.5	Other ferro-alloys	18,364	7.0
	678.4	Hydro-electric conduits	228	10.0

Notes: See notes to Table 4.A.4.

(discussed in the last section).

Details of the commodities covered, together with the relevant 1965/67 tariffs, are shown in Table 4.E.1, and the two frequency distributions of tariff rates in Chart 4.E.1. The distributions are rather flat in shape, a reflection of the low level of tariff protection on most of the non-ferrous metal items and the rather higher tariffs imposed on the metal manufactures. Tariff differences cover quite a wide range as Chart 4.E.2 indicates, but the incidence of extreme cases is small and for well over half the products differences between UK and EEC tariffs lie in the 0 to +4 bracket. Products for which UK and EEC tariffs were similar, or on which the EEC imposed a higher duty than the UK did, are, with one exception, all non-ferrous metals, whilst the items referred to on the right of the graph, i.e., products on which the UK levied a higher rate of duty, are predominantly metal manufactures.

Aggregate trade data are shown in Table 4.E.2. British exports to the EEC came to nearly $270 million, whilst EEC exports to the UK were less than half that figure – $130 million. Third country exports were about $430 million for the

UK and $880 million for the EEC. The neutral market export ratio, at 49.3, works out at a little above that for total trade, which would suggest an overall comparative cost position for the present sector just slightly better than the average, whereas the crude trade balance signals a rather more promising outlook. The ratios in column 6 are in a continually descending order and thus indicate a good measure of agreement between the two performance tests.

Details of UK and EEC export performances are shown in Table 4.E.3. A first and obvious point to note is that classes I and II, with 15 out of 24 items assigned to them, are very much larger than the remaining classes. On the other hand, the five conflicting evidence products shown in the table are all to be found in the first two classes. Although the data underlying the mutual trade ranking order take account of tariffs whereas the trade statistics shown in Table 4.E.2 do not, the detailed results thus bear out the earlier finding that the mutual trade figures give a rather better impression of the overall performance of this industry section than third country exports do.

When classes I and II are considered in a little more detail it is interesting to see that class I is almost entirely composed of non-ferrous metals whereas class II largely consists of metal manufactures. The position is particularly clear-cut without the conflicting evidence cases – in class I, Worked Copper, Unwrought Lead, and Nuts and Bolts, in class II, Worked Lead and Miscellaneous Non-Ferrous Metals; if these items are left out of account, class I consists of five non-ferrous metals and class II of five metal manufactures.

No special comment appears to be called for with regard to class III. The composition of class IV is noteworthy because of the agreed assignment to that class of both aluminium items (Unwrought Aluminium and Worked Aluminium). Class V contains only one product, which, moreover, is quantitatively one of the least important – Nails etc.

Before the discussion can move on some thought should be given to the interpretation of the results for non-ferrous metals, particularly in the unwrought state. Taken at their face value the figures in Table 4.A.3 would suggest that in an enlarged EEC British producers of, say, Unwrought Nickel or Unwrought Lead would be in a comparatively strong position vis-a-vis their competitors in the continental member countries, whereas the position would be reversed with regard to, say, Unwrought Zinc. Two factors combine to lead one to question how far this is a realistic approach. In the first place, UK and/or EEC tariffs on several of the items concerned are already zero; in these cases UK accession to the EEC would not lead to any obvious changes in the relative positions of UK and EEC producers in each other's markets. Secondly, in view of the fact that neither the UK nor the EEC mines more than very limited quantities of non-ferrous metal ores it is not clear in what sense Unwrought Lead, for example, can be said to be 'produced' in the UK for export to the EEC, especially as imports of the ores or concentrates from which the metals could be smelted often turn out to be quite insignificant. One fairly obvious possibility – that trade between the UK and the EEC consists of re-exports – is ruled out by the nature of the trade statistics used, for re-exports are excluded throughout. Another possibility is that mutual UK/EEC trade largely takes the form of trade in alloys which, as Table 4.E.1 shows, are generally part of the same trade item as the pure metal. Detailed knowledge of the nature of trade in non-ferrous metals is required to throw light on this point and in general to make sense of the data; in the absence of such knowledge we can go no further than to suggest that the results for the unwrought metals may well be rather misleading.[15]

The final topic is trade diversion. Table 4.E.4 provides the usual set of data – the EEC's third country imports of class I and II products in the upper part of the table, the UK's imports of class IV and V items in the lower part, in each case together with the relevant rates of the post-Kennedy Round rates of the CET. *Prima facie* opportunities for UK exporters would appear to be promising with regard to a number of products – Unworked Silver, Unworked Platinum, Other Tools, Other Metal Manufactures, and two conflicting evidence products, Worked Copper, and Miscellaneous Non-Ferrous Metals. However, the first two of these are among the commodities for which the findings of Table 4.E.3 may be misleading, and it is therefore questionable how far any trade diversion analysis can be applied to them. Furthermore, the conflict of evidence regarding the last two products means that there is some doubt about the ability of the British producers concerned to exploit opportunities for replacing imports from third countries. In the end the list is thus reduced to only two firm items – Other Tools and Other Metal Manufactures.

The outlook for EEC producers is equally limited. The lower half of the table contains only one product which the UK purchased on a large scale from third countries and on which the CET is not zero – Unwrought Aluminium; in addition, this is, of course, one of the products for which our results may not be valid.

F. Mechanical Engineering

The present section is the first of three dealing with what some readers may regard as the central issue of the study – the effect of UK accession to the EEC upon the engineering industry, in rather a wide sense of the term. The discussion begins with mechanical engineering (described in the SITC as Machinery, other than Electric).

Table 4.F.1 shows details of the 29 products covered and also sets out the tariffs which were imposed by the UK and the EEC respectively in 1965/67. As usual the tariff data are considered in two stages – first with reference to the frequency distributions of tariff rates (Chart 4.F.1), and subsequently with reference to the differences between UK and EEC tariffs on the same products (Chart 4.F.2). The first chart shows both distributions to be markedly peaked; the EEC's rates are clustered in the 10 to 14.9 percentage points bracket, whilst the main concentration of UK tariffs occurs in the 15 to 19.9 points range. The distribution of tariff differences (Chart 4.F.2) calls for two comments – (a) tariff differences for the 29 products are fairly similar, as apart from the two items referred to in the +10 bracket (Internal Combustion Engines and Machine Tools) the differences extend over only four intervals, and (b) there is not a single

15 At an early stage in the work for the study we considered the possibility of altogether excluding some or all unwrought metals, for the reasons given in the text. We decided against this; the objections voiced probably apply rather less strongly to some items (aluminium?) than to others, and it would have been difficult to justify drawing a line between products to be omitted and those to be retained.

Table 4.E.1 *Non-Ferrous Metals, etc.: Products Covered and Tariff Rates*

SITC	Description	Tariffs (1965/67; per cent) UK	EEC	UK-EEC
681.1	Silver, unworked or partly worked	7.5	9.3	−1.8
681.2	Platinum and other metals of the platinum group, unworked or partly worked	6.5	5.5	1.0
682.1	Copper and alloys, whether or not refined, unwrought	6.8	0	6.8
682.2	Copper and alloys of copper, worked	17.0	10.0	7.0
683.1	Nickel and nickel alloys, unwrought	0	0	–
683.2	Nickel and nickel alloys, worked	8.0	7.2	0.8
684.1	Aluminium and aluminium alloys, unwrought	0	2.5	−2.5
684.2	Aluminium and aluminium alloys, worked	15.0	15.0	–
685.1	Lead and lead alloys, unwrought	0	0	–
685.2	Lead and lead alloys, worked	3.5	7.0	−3.5
686.1	Zinc and zinc alloys, unwrought	10.0	0	10.0
686.2	Zinc and zinc alloys, worked	13.5	10.0	3.5
687.1	Tin and tin alloys, unwrought	0	0	–
689	Miscellaneous non-ferrous base metals employed in metallurgy	11.0	6.5	4.5
691.1	Finished structural parts and structures of iron or steel	14.0	14.0	–
692.1	Tanks, vats and reservoirs for storage or manufacturing use	19.0	13.7	5.3
693.1	Wire cables, ropes, plaited bands, slings and similar articles, not insulated	21.0	15.0	6.0
694.1	Nails, tacks, staples, spikes, etc.	16.0	12.3	3.7
694.2	Nuts, bolts, screws, rivets, washers, etc.	16.7	13.7	3.0
695.2	Other tools for use in the hand or in machines	16.0	12.0	4.0
696.0	Cutlery	19.0	15.0	4.0
697.1	Domestic stoves, boilers, cookers, ovens, space heaters, n.e.s.	18.7	13.5	5.2
697.2	Domestic utensils of base metals	18.0	15.5	2.5
698	Manufactures of metal such as locksmiths' wares, safes, chain, anchors, pins and needles, springs, etc.	18.9	15.4	3.5

Notes: See Table 4.A.1.

Chart 4.E.1 *UK and EEC Tariff Rates on Non-Ferrous Metals, etc.* (1965/67)

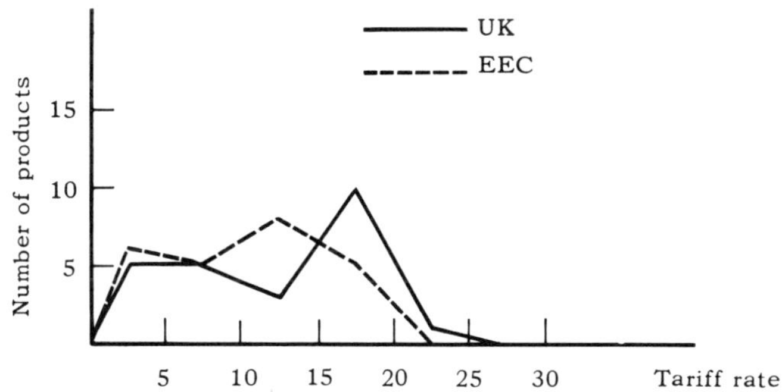

Source: Table 4.E.1.

Chart 4.E.2 *Differences between Tariffs on Non-Ferrous Metals, etc.* (UK minus EEC, 1965/67)

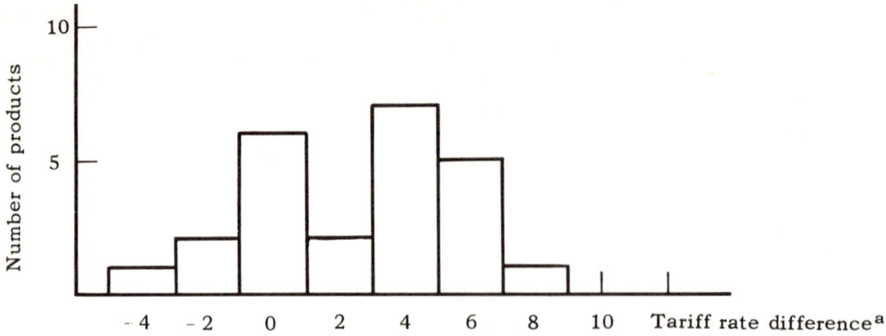

[a]See note (a) for Chart 4.A.2

Source: Table 4.E.1.

Table 4.E.2 *Trade in Non-Ferrous Metals, etc. by Class (Mutual Trade Ranking)* (1965/67 average)

Class	No. of items	UK exports to EEC ($million) (1)	EEC exports to UK ($million) (2)	Intra-trade ($million) (3)	UK exports to third countries ($million) (4)	EEC exports to third countries ($million) (5)	(4)/(5) (per cent) (6)
I	8	131.4	16.8	466.1	150.8	198.5	76.0
II	7	61.9	38.1	531.9	201.5	404.4	49.8
III	4	67.5	53.0	455.1	45.3	132.8	34.1
IV	4	6.8	19.7	254.4	33.3	128.4	25.9
V	1	0.6	2.0	9.7	2.6	14.4	17.9
Total	24	268.3	129.6	1,717.1	433.4	878.4	49.3
Non-ferrous metals as % of all manufactures		13.2	6.9	8.9	6.7	6.4	–

Notes: See notes (i) to (v) of Table 4.A.3.

Table 4.E.3 *Trade in Non-Ferrous Metals, etc.: A Detailed Comparison (1965/67 average)*

Class	SITC	Description	UK exports to EEC ($'000) (1)	EEC exports to UK ($'000) (2)	Intra-trade ($'000) (3)	UK exports to third countries ($'000) (4)	EEC exports to third countries ($'000) (5)	Special features (6)
I	681.1	Silver, unworked	29,817	2,725	62,696	3,594	3,758	
	681.2	Platinum, unworked	22,943	3,207	43,341	54,799	13,514	
	682.2	Copper, worked	13,247	3,336	154,140	45,871	124,726	* III
	683.1	Nickel, unwrought	33,105	776	39,841	3,443	5,998	
	683.2	Nickel, worked	7,724	690	19,608	7,704	8,404	
	685.1	Lead, unwrought	9,980	851	31,690	5,128	11,574	III
	687.1	Tin, unwrought	8,225	3,000	64,851	19,716	4,679	*
	694.2	Nuts & bolts	6,325	2,220	49,946	10,541	25,845	III
II	685.2	Lead, worked	80	4	2,648	573	1,991	IV
	689	Miscellaneous non-ferrous metals	6,892	4,235	37,604	7,186	29,197	IV
	691.1	Structures & parts	5,176	2,302	79,799	37,428	93,856	
	693.1	Wire cables, etc.	2,140	1,178	25,288	18,620	33,971	
	695.2	Other tools	16,737	12,458	117,998	46,530	86,186	
	696.0	Cutlery	4,733	3,073	42,967	25,998	32,874	
	698	Other metal manufactures	26,190	14,839	225,595	65,132	126,294	
III	682.1	Copper, unwrought	64,165	49,791	365,663	28,829	97,727	
	686.2	Zinc, worked	237	220	7,181	1,115	1,717	
	697.1	Stoves & cookers	1,190	1,496	50,756	5,571	16,716	
	697.2	Domestic utensils	1,945	1,496	31,479	9,739	16,675	
IV	684.1	Aluminium, unwrought	1,850	4,568	86,080	5,838	18,811	
	684.2	Aluminium, worked	4,347	11,266	117,851	22,291	76,876	
	686.1	Zinc, unwrought	184	2,506	37,550	651	10,178	
	692.1	Tanks, etc.	452	1,337	12,877	4,504	22,510	
V	694.1	Nails, etc.	586	2,045	9,701	2,567	14,363	

Notes: See notes to Table 4.A.3.

Table 4.E.4 *Non-Ferrous Metals: Determinants of Trade Diversion*

(a) In favour of UK sources of supply

Class	SITC	Description	EEC imports from third countries (1965/67 average; $'000)	Post-Kennedy CET (per cent)
I	681.1	Silver, unworked	86,506	4.7
	681.2	Platinum, unworked	33,104	2.8
	682.2	*Copper, worked*	27,267	7.5
	683.1	Nickel, unwrought	34,266	0
	683.2	Nickel, worked	7,021	5.2
	685.1	*Lead, unwrought*	39,979	0
	687.1	Tin, unwrought	63,516	0
	694.2	*Nuts & bolts*	12,845	7.4
II	685.2	*Lead, worked*	1,005	9.3
	689	*Miscellaneous non-ferrous metals*	69,763	7.8
	691.1	Structures & parts	3,015	5.5
	693.1	Wire cables, etc.	1,273	9.5
	695.2	Other tools	35,130	6.6
	696.0	Cutlery	10,870	11.1
	698	Other metal manufactures	34,716	7.5

(b) In favour of EEC sources of supply

Class	SITC	Description	UK imports from third countries (1965/67 average; $'000)	Post-Kennedy CET (per cent)
IV	684.1	Aluminium, unwrought	114,362	1.3
	684.2	Aluminium, worked	5,833	10.9
	686.1	Zinc, unwrought	51,744	0
	692.1	Tanks, etc.	1,058	6.8
V	694.1	Nails, etc.	532	6.1

Notes: See notes to Table 4.A.4.

product among the 29 for which the EEC tariff exceeded the UK rate of duty. Inspection of Table 4.F.1 suggests that there are no very obvious associations between particular groups of products on the one hand and either the tariffs themselves or the tariff differences on the other.

The size of aggregate trade flows (both mutual and in the form of exports to third countries) is much the largest for any of the nine industry sections. As Table 4.F.2 shows, UK exports to the EEC averaged $540 million and EEC exports to the UK came to only slightly less that amount; in the case of exports to third countries the rounded figures are $1,700 million for the UK and $3,000 million for the EEC. The overall ratio of UK and EEC neutral market exports works out at nearly 57 and thus indicates a rather good performance by British producers, but the crude trade balance, only marginally in the UK's favour, gives less ground for optimism. Column 6 shows that agreement between the two performance tests is fairly good, the main irregularity being that the third country export ratios for classes II and III are in inverse order, by quite a big margin.

Details of the UK's and the EEC's export performances in mechanical engineering products are set out in Table 4.F.3. Only three items are placed in class I, but on the other hand the high rank is not disputed by the evidence of neutral market exports; indeed, two of the products (Aircraft Engines and Tractors) are distinguished by particularly strong third country export records. Class II contains five entries, two of which (Other Metal Working Machinery, and Non-Electrical Domestic Appliances) perform much less well in the third country export comparison, with the result that the alternative rank order assigns them to class IV. On the other hand, both of these, and especially the last one, are quantitatively much less important than the remaining three – Pumps etc., Mechanical Handling Equipment, and Machinery Parts (a class I/II borderline case) – whose classing is not contested. Class III is quite small, with only three products assigned to it. On the evidence of neutral market exports two of these - Cultivating Machinery and Other Office Machinery – would move to class I.

In contrast to the first three groups of products the fourth class is very large, with 12 items assigned to it. From the point of view of the UK this is clearly rather a disturbing feature, though the position looks a little better in terms of neutral market exports, as the alternative ranking would place three commodities – Textile Machinery, Sewing Machines, and Ball Bearings – in class II. Quantitatively important items whose assignment to class IV is not contested are Harvesters, Printing Machinery, Heating and Cooling Equipment, Other Non-electrical Machinery, and Other Machines and Mechanical Appliances.

Class V presents a rather untidy picture, as four out of the six constituents are conflicting evidence cases; three items – Machine Tools, Food Processing Machines, and Machinery for Mineral Products – are placed by the alternative performance test in class III, the fourth, Steam Engines (quantitatively of minor importance) right at the other end of the scale in class I. Of these results the most significant is that for Machine Tools; it is clearly a serious matter that this item does not get above class III in either rank order.

Attention should also be paid to the two products whose assignment to class V is not contested – Typewriters and Statistical Machines. Typewriters are noteworthy simply because of the abysmal performance of the item, both in mutual trade and with respect to third country exports. Statistical Machines, on the other hand, may be singled out because this is one of the two headings under which computers are classified in the trade returns. As the other item covering computers (Calculating Machines[16]) is uncontestedly placed in class IV, the results point to a rather pessimistic conclusion concerning the relative position of the UK computer industry in an enlarged EEC. (It must be remembered, however, that both 'Statistical Machines' and 'Calculating Machines' cover items other than what may be called computers in the narrow sense of the word, and that the evidence is therefore by no means definitive even within the framework of the present analysis.)

By way of conclusion it may be worth pointing out that the products under discussion on the whole perform rather better in the third country export test than

16 The detailed descriptions of commodities in the British trade statistics – which are based on the SITC – make it appear that digital computers are classified as Calculating Machines (SITC 714.2), while punched card computers rank as Statistical Machines (SITC 714.3).

they do in the mutual trade comparison. In the first place, of the 11 contested items only two move down in the rank order on the basis of the neutral market export record, and the other 9 move to higher classes. Secondly, it so happens that there are a relatively large number of cases of products being placed by the third country export test in a class just one above that to which they are assigned by the mutual trade ranking. This difference in class is not, of course, sufficient to turn the products into conflicting evidence items, but nonetheless is worthy of mention. The number of products which the neutral market export rank order would *in toto* assign to each class is as follows:

class I – 6
class II – 6
class III – 12
class IV – 4
class V – 1

Discussion of opportunities for trade diversion is based on the data assembled in Table 4.F.4. The upper part of the table shows a number of items which the EEC imported on a substantial scale from third countries and which *prima facie* afford the UK producers concerned good scope for diverting trade in their favour – all three constituents of class I, Machinery Parts, Pumps etc., and Mechanical Handling Equipment out of class II, and possibly Other Office Machinery from class III (a conflicting evidence item). However, it should perhaps be added that markets for two of the products referred to – Aircraft Engines and Machinery Parts – are probably rather imperfect (though for somewhat different reasons), so that substitution of British for third country products may face particular difficulty in these cases.[17]

By way of contrast British third country imports of class IV and V products, shown in the lower part of the table, were mostly relatively small. However, two of the three items which may be singled out as quantitatively important are conflicting evidence products (Machine Tools, and Textile Machinery) so that it is not clear that EEC producers would be sufficiently strong to exploit opportunities for diverting trade away from third country suppliers on the scale indicated. (If the neutral market export evidence turned out to be the more reliable test, one of these – Textile Machinery – would join the items which may afford sizeable trade diversion gains to British producers.) There is thus only one product out of the 18 in classes IV and V – Statistical Machines – in which EEC producers may expect significant trade diversion benefits, though in combination the remaining 17 do, of course, add up to a large total.

17 Moreover, in the past the provisions concerning the CET on Aircraft Engines have been such that in practice most EEC imports have probably been free of duty. Unless the rules are altered British manufacturers may therefore not gain any advantages relative to outside producers.

Table 4.F.1 *Mechanical Engineering: Products Covered and Tariff Rates*

SITC	Description	Tariffs (1965/67; per cent) UK	EEC	UK-EEC
711.3	Steam engines and turbines	17.5	11.0	6.5
711.4	Aircraft engines including jet propulsion engines	15.3	10.5	4.8
711.5	Internal combustion engines, other than for aircraft	24.6	14.0	10.6
712.1	Agricultural machinery and appliances for preparing and cultivating the soil	12.0	9.0	3.0
712.2	Agricultural machinery and appliances for harvesting, threshing and sorting ('Harvesters')	12.3	9.0	3.3
712.5	Tractors, other than road tractors for tractor-trailer combinations	17.5	15.0	2.5
714.1	Typewriters and cheque-writing machines	14.2	11.5	2.7
714.2	Calculating machines, accounting machines and similar machines incorporating a calculating device	13.0	11.7	1.3
714.3	Statistical machines, e.g., calculating from punched cards or tape	14.0	9.0	5.0
714.9	Other office machinery such as duplicating and addressing machines etc. and office machinery parts	20.3	12.2	8.1
715.1	Machine-tools for working metals	17.5	6.7	10.8
715.2	Metal working machinery, other than machine-tools	15.3	10.2	5.1
717.1	Textile machinery	15.6	11.0	4.6
717.2	Machinery for preparing, tanning or working hides, skins or leather	11.0	10.5	0.5
717.3	Sewing machines	16.0	13.0	3.0
718.1	Paper mill and pulp mill machinery and other machinery for the manufacture of paper articles	13.5	10.4	3.1
718.2	Printing and bookbinding machinery	12.3	10.3	2.0
718.3	Food-processing machines (excluding domestic)	15.0	13.0	2.0
718.4	Construction and mining machinery such as road rollers, excavating, levelling, boring, etc. machinery	15.5	10.7	4.8
718.5	Mineral crushing, sorting and moulding machinery; glass working machinery	13.3	9.8	3.5
719.1	Heating and cooling equipment	14.5	11.0	3.5
719.2	Pumps and centrifuges	16.2	12.1	4.1
719.3	Mechanical handling equipment	15.5	13.2	2.3
719.4	Domestic appliances (non-electrical)	14.0	11.7	2.3
719.5	Other powered-tools such as machine-tools for working minerals, wood, plastics, etc.	15.9	9.8	6.1
719.6	Other non-electrical machines such as calendering machines, weighing machinery, spraying machinery and automatic vending machines	16.3	11.5	4.8
719.7	Ball, roller or needle-roller bearings	18.0	18.0	–
719.8	Machinery and mechanical appliances, n.e.s.	16.6	12.5	4.1
719.9	Parts and accessories of machinery, n.e.s.	16.5	11.4	5.1

Notes: See Table 4.A.1.

Chart 4.F.1 *UK and EEC Tariff Rates on Mechanical Engineering Products* (1965/67)

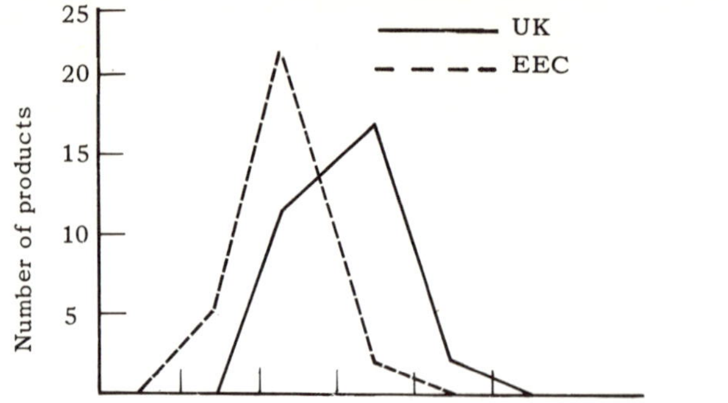

Source: Table 4.F.1.

Chart 4.F.2 *Differences between Tariffs on Mechanical Engineering Products* (UK minus EEC, 1965/67)

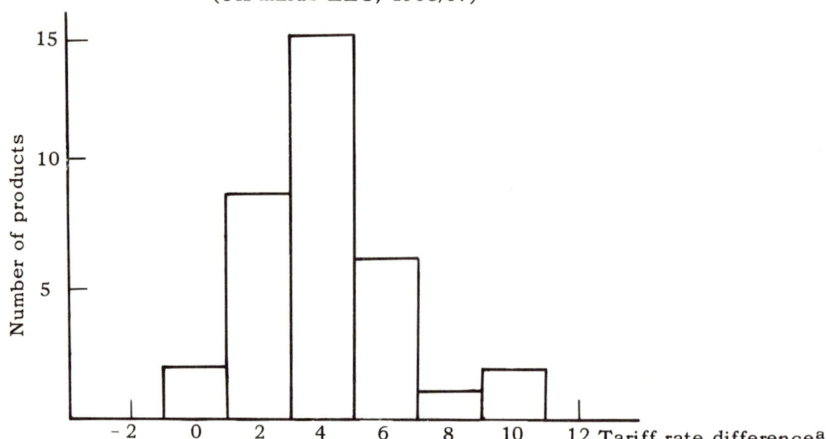

[a]See note (a) for Chart 4.A.2.

Source: Table 4.F.1.

Table 4.F.2 *Trade in Mechanical Engineering Products by Class (Mutual Trade Ranking)* (1965/67 average)

Class	No. of items	UK exports to EEC ($million) (1)	EEC exports to UK ($million) (2)	Intra-trade ($million) (3)	UK exports to third countries ($million) (4)	EEC exports to third countries ($million) (5)	(4)/(5) (per cent) (6)
I	3	140.6	56.7	541.5	502.6	281.8	178.4
II	5	119.6	77.2	824.7	310.7	590.4	52.6
III	3	63.0	50.9	333.9	126.5	189.7	66.7
IV	12	161.2	205.9	1,367.0	584.1	1,441.5	40.5
V	6	56.6	140.8	623.9	210.0	545.8	38.5
Total	29	540.9	531.5	3,691.0	1,734.0	3,049.2	56.9
Mechanical engineering products as % of all manufactures		26.5	28.4	19.3	26.6	22.3	–

Notes: See notes (i) to (v) of Table 4.A.3.

Table 4.F.3 *Trade in Mechanical Engineering Products: A Detailed Comparison*

(1965/67 average)

Class	SITC	Description	UK exports to EEC ($'000) (1)	EEC exports to UK ($'000) (2)	Intra-trade ($'000) (3)	UK exports to third countries ($'000) (4)	EEC exports to third countries ($'000) (5)	Special features (6)
I	711.4	Aircraft engines	52,732	35,450	132,837	120,037	34,631	
	711.5	Other int. comb. engines	63,716	17,893	269,210	196,046	184,062	
	712.5	Tractors	24,138	3,361	139,473	186,498	63,075	
II	715.2	Other metal working mach.	7,570	5,092	64,273	20,432	64,299	IV
	719.2	Pumps & centrifuges	36,359	25,063	240,890	91,372	208,980	
	719.3	Mechanical handling equip.	22,961	18,678	189,671	86,945	139,550	*
	719.4	Non-elec. dom. appliances	1,418	570	18,281	2,511	8,226	IV
	719.9	Machinery parts	51,244	27,768	311,624	109,485	169,368	†
III	712.1	Cultivating machinery	1,821	2,312	21,863	14,362	8,588	* I
	714.9	Other office machinery	36,865	27,384	177,158	32,848	39,956	I
	718.4	Construction machinery	24,328	21,212	134,869	79,316	141,189	
IV	712.2	Harvesters	15,681	17,666	135,332	14,443	33,757	
	714.2	Calculating machines	7,769	10,942	75,902	39,520	106,820	
	717.1	Textile machinery	37,582	39,476	231,794	149,531	322,566	II
	717.2	Leather working machines	1,415	1,713	10,436	6,924	17,320	
	717.3	Sewing machines	5,299	7,193	48,601	24,749	54,821	II
	718.1	Paper machinery	6,560	9,474	61,598	27,111	65,139	
	718.2	Printing machinery	14,803	21,811	103,572	33,312	99,772	*
	719.1	Heating & cooling equip.	18,467	22,246	204,531	81,856	213,690	
	719.5	Other power tools	9,031	14,767	97,902	33,426	96,208	
	719.6	Other non-electrical mach.	12,498	15,916	122,379	35,119	97,824	
	719.7	Ball bearings	6,451	7,662	58,074	25,332	47,429	† II
	719.8	Other machinery	25,661	37,036	216,895	112,774	286,195	
V	711.3	Steam engines	939	4,270	22,807	31,487	21,930	I
	714.1	Typewriters	662	16,935	56,304	8,173	68,271	
	714.3	Stat. machines	27,822	45,801	210,954	12,387	43,571	
	715.1	Machine tools	21,260	58,206	237,490	92,457	261,948	III
	718.3	Food processing machines	2,148	6,904	34,943	34,002	78,005	III
	718.5	Mach. for mineral products	3,747	8,691	61,429	31,513	72,036	III

Notes: See notes to Table 4.A.3.

Table 4.F.4 *Mechanical Engineering Products: Determinants of Trade Diversion*

(a) In favour of UK sources of supply

Class	SITC	Description	EEC imports from third countries (1965/67 average; $'000)	Post-Kennedy CET (per cent)
I	711.4	Aircraft engines	70,630	7.9
	711.5	Other internal combustion engines	51,516	8.5
	712.5	Tractors	23,308	7.5
II	715.2	*Other metal working machinery*	6,591	5.4
	719.2	Pumps & centrifuges	60,166	6.1
	719.3	Mechanical handling equipment	44,713	6.6
	719.4	*Non-electrical domestic appliances*	1,395	5.8
	719.9	Machinery parts	80,657	5.8
III	712.1	*Cultivating machinery*	1,525	4.5
	714.9	*Other office machinery*	72,363	6.6

(b) In favour of EEC sources of supply

Class	SITC	Description	UK imports from third countries (1965/67 average; $'000)	Post-Kennedy CET (per cent)
IV	712.2	Harvesters	3,272	4.5
	714.2	Calculating machines	13,492	7.0
	717.1	*Textile machinery*	24,384	4.5
	717.2	Leather working machines	440	4.5
	717.3	*Sewing machines*	9,699	7.3
	718.1	Paper machinery	4,643	5.2
	718.2	Printing machinery	9,568	5.0
	719.1	Heating & cooling equipment	14,497	5.6
	719.5	Other power tools	16,358	6.0
	719.6	Other non-electrical machinery	12,764	5.8
	719.7	*Ball bearings*	10,882	9.0
	719.8	Other machinery	19,611	6.0
V	711.3	*Steam engines*	2,038	5.8
	714.1	Typewriters	3,349	5.7
	714.3	Statistical machines	30,129	7.0
	715.1	*Machine tools*	41,389	6.3
	718.3	*Food processing machines*	2,830	5.8
	718.5	*Machinery for mineral products*	3,802	4.8

Notes: See notes to Table 4.A.4.

G. Electrical Engineering

The second section dealing with the engineering industry is devoted to electrical goods and some associated products. The number of commodities included is relatively small – 19; detailed descriptions of the goods concerned, together with the appropriate rates of duty prescribed in the UK and EEC tariff lists, are set out in Table 4.G.1.

The frequency distributions of the two sets of tariffs are shown in Chart 4.G.1. The main point to note is clearly that the EEC distribution has a peak in the 10 to 14.9 percentage points interval, whereas the distribution for the UK is quite flat, with an equal number of tariff rates in the 15–19.9 and 20–24.9 points brackets. In view of the lack of similarity between the two distributions it is not perhaps surprising that tariff differences extend over a wide range, as is shown in Chart 4.G.2, though for the majority of the 19 items tariff differences are nonetheless confined to the 0 to +4 intervals. Examination of Table 4.G.1 suggests that as in the case of mechanical engineering neither the two sets of tariffs themselves, nor the tariff differences, appear to follow any kind of regular pattern.

Data of mutual and third country exports, in the aggregate and by performance class, are shown in Table 4.G.2. UK exports to the EEC averaged $165 million, whereas trade in the opposite direction came to a slightly lower figure – $156 million. UK and EEC exports to third countries amounted to just over $680 million and nearly $1,460 million respectively. The crude trade balance (which was just positive) and the overall neutral market export ratio (which works out at 46.7 – just a little below the average) both show that the industry as a whole compares neither particularly favourably nor particularly unfavourably with the remainder of the UK manufacturing sector. The ratios in column 6 follow a somewhat erratic course and thus do not indicate any very close agreement between the two performance measures at the class level.

Detailed export figures for the 19 commodities covered are set out in Table 4.G.3. Class I contains four entries (including one conflicting evidence product, Electric Equipment for Vehicles) all of them constituents of SITC 729 – Other Electrical Machinery and Apparatus. Quantitatively much the most important item is Electrical Measuring and Control Apparatus.

Only two products are assigned to class II. One of these is the quantitatively important Other Telecommunications Equipment item, which covers telecommunications apparatus other than television sets and radios. With regard to class III (five entries) particular attention should perhaps be drawn to the agreed assignment of both constituents of SITC 722, which covers 'heavy' electrical equipment.

Class IV also contains five items. The neutral market export rank order would place two of these – Insulating Equipment and Television Sets – in class I, but neither of them is quantitatively of special significance. It should be added that in the case of Radio Sets – the other telecommunications equipment commodity in class IV - the UK's third country export record is particularly poor. Only three products are assigned to class V (including one conflicting evidence good – Gramophones); quantitatively the most important of these is the Valves and Tubes item.

The question of trade diversion can be disposed of quite quickly. The figures in the upper part of Table 4.G.4 (which contains the usual information – the EEC's third country imports of class I and II products and EEC post-Kennedy Round tariffs) show that there are two items – Electrical Measuring and Control Apparatus, and Other Telecommunications Equipment – which the EEC imported on a substantial scale from third countries and which would therefore afford the relevant UK exporters scope for raising exports at the expense of third country producers by large margins. Opportunities open to EEC suppliers are rather more limited. As the lower part of the table indicates, there is only one class IV/V product which the UK purchased in quantity from third countries – Valves & Tubes; moreover, the post-Kennedy Round CET rate is not significantly higher than average.

Table 4.G.1 *Electrical Engineering: Products Covered and Tariff Rates*

SITC	Description	Tariffs (1965/67; per cent) UK	EEC	UK-EEC
722.1	Electric power machinery	13.5	12.0	1.5
722.2	Electrical apparatus for making and breaking electrical circuits (switchgears, etc.)	24.3	12.6	11.7
723.1	Insulated wire and cable	10.0	15.5	−5.5
723.2	Electrical insulating equipment	17.5	16.0	1.5
724.1	Television broadcast receivers	24.0	22.0	2.0
724.2	Radio broadcast receivers	20.0	22.0	−2.0
724.9	Other telecommunications equipment such as telephone apparatus, microphones and amplifiers	18.5	14.3	4.2
725.0	Domestic electrical equipment	14.5	13.1	1.4
726	Electric apparatus for medical purposes and radiological apparatus	26.0	13.0	13.0
729.1	Batteries and accumulators	19.0	18.4	0.6
729.2	Electric lamps	22.0	13.0	9.0
729.3	Thermionic, etc. valves and tubes, photocells, transistors, etc.	20.5	14.7	5.8
729.4	Automotive electrical equipment ('Electric equipment for vehicles')	23.1	14.5	8.6
729.5	Electrical measuring and controlling instruments and apparatus	22.0	12.5	9.5
729.6	Electro-mechanical hand tools	14.0	14.0	–
729.9	Electrical machinery and apparatus, n.e.s.	18.3	12.4	5.9
812.4	Lighting fixtures and fittings, lamps and lanterns	17.7	16.5	1.2
891.1	Gramophones, tape recorders, and other sound recorders and reproducers	17.7	13.8	3.9
891.2	Gramophone records, recorded tapes, etc.	16.7	13.4	3.3

Notes: See Table 4.A.1.

Chart 4.G.1 *UK and EEC Tariff Rates on Electrical Engineering Products* (1965/67)

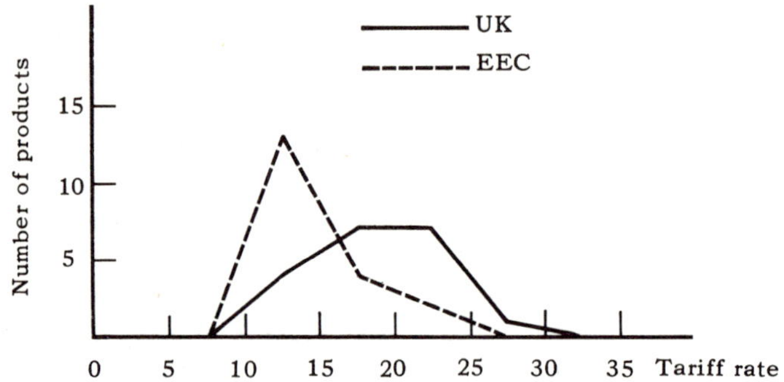

Source: Table 4.G.1.

Chart 4.G.2 *Differences between Tariffs on Electrical Engineering Products* (UK minus EEC, 1965/67)

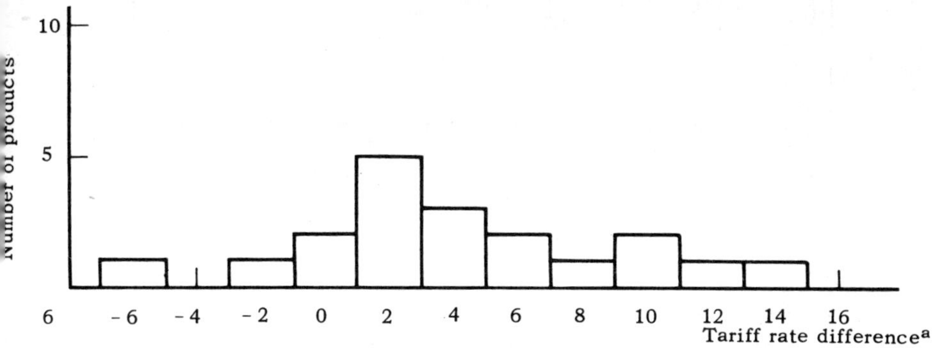

[a]See note (a) for Chart 4.A.2.
Source: Table 4.G.1.

Table 4.G.2 *Trade in Electrical Engineering Products by Class (Mutual Trade Ranking)* (1965/67 average)

Class	No. of items	UK exports to EEC ($million) (1)	EEC exports to UK ($million) (2)	Intra-trade ($million) (3)	UK exports to third countries ($million) (4)	EEC exports to third countries ($million) (5)	(4)/(5) (per cent) (6)
I	4	49.7	21.1	241.5	82.6	147.1	56.2
II	2	26.8	18.4	275.7	222.4	342.5	64.9
III	5	67.1	63.2	810.9	289.2	645.9	44.8
IV	5	7.7	18.2	271.0	39.3	150.2	26.2
V	3	14.0	35.2	219.5	47.5	174.1	27.3
Total	19	165.4	156.2	1,818.5	681.1	1,459.8	46.7
Electrical engineering products as % of all manufactures		8.1	8.4	9.5	10.5	10.7	–

Notes: See notes (i) to (v) of Table 4.A.3.

Table 4.G.3 *Trade in Electrical Engineering Products: A Detailed Comparison*

(1965/67 average)

Class	SITC	Description	UK exports to EEC ($'000) (1)	EEC exports to UK ($'000) (2)	Intra-trade ($'000) (3)	UK exports to third countries ($'000) (4)	EEC exports to third countries ($'000) (5)	Special features (6)
I	729.1	Batteries, etc.	6,590	1,457	29,709	25,357	25,808	
	729.4	Elect. equip. for vehicles	9,498	3,790	60,173	13,717	36,676	III
	729.5	Elect. control apparatus	27,873	14,265	127,480	38,008	75,234	
	729.6	Electric hand tools	5,748	1,602	24,187	5,514	9,352	
II	723.1	Insulated wire	3,741	2,186	62,594	72,944	87,010	
	724.9	Other telecom. equipment	23,053	16,255	213,078	149,445	255,523	
III	722.1	Elect. power machinery	15,016	12,052	175,869	119,046	273,185	†
	722.2	Switchgear, etc.	15,814	14,690	217,397	80,376	145,911	
	725.0	Domestic elect. equip.	17,298	17,912	264,613	40,726	92,996	
	729.9	Other elect. machinery	15,946	15,090	128,399	42,967	118,906	
	891.2	Records & tapes	3,049	3,500	24,573	6,118	14,883	
IV	723.2	Insulating equipment	823	2,142	16,641	6,204	8,404	I
	724.1	Television sets	205	1,346	30,138	5,471	4,888	I
	724.2	Radio sets	1,472	4,275	102,198	4,381	65,566	
	729.2	Electric lamps	3,623	6,539	75,707	10,572	33,173	
	812.4	Lighting fixtures	1,561	3,870	46,277	12,697	38,208	
V	726	Elect. medical apparatus	1,181	5,100	31,587	3,441	28,384	
	729.3	Valves & tubes	9,362	21,516	131,319	16,626	82,251	
	891.1	Gramophones, etc.	3,505	8,620	56,569	27,478	63,472	III

Notes: See notes to Table 4.A.3.

Table 4.G.4 *Electrical Engineering Products: Determinants of Trade Diversion*

(a) In favour of UK sources of supply

Class	SITC	Description	EEC imports from third countries (1965/67 average; $'000)	Post-Kennedy CET (per cent)
I	729.1	Batteries, etc.	7,860	13.9
	729.4	*Elect. equipment for vehicles*	10,272	8.5
	729.5	Elect. control apparatus	89,801	9.4
	729.6	Electric hand tools	6,720	7.0
II	723.1	Insulated wire	10,377	11.0
	724.9	Other telecom. equipment	60,993	7.8

(b) In favour of EEC sources of supply

Class	SITC	Description	UK imports from third countries (1965/67 average; $'000)	Post-Kennedy CET (per cent)
IV	723.2	*Insulating equipment*	4,357	9.5
	724.1	*Television sets*	532	14.0
	724.2	Radio sets	11,994	14.0
	729.2	Electric lamps	2,462	6.9
	812.4	Lighting fixtures	3,280	9.0
V	726	Elect. medical apparatus	1,992	7.3
	729.3	Valves & tubes	26,850	10.0
	891.1	*Gramophones, etc.*	12,437	8.1

Notes: See notes to Table 4.A.4.

H. Transport Equipment

The third sub-division of the engineering sector takes the discussion to the producers of transport equipment – the car industry, the aircraft industry, and shipbuilding. The number of transport equipment items separately distinguished is 14, which makes this much the smallest of the nine industry sections.

Details of the commodities included, together with the appropriate UK and EEC tariff rates, can be found in Table 4.H.1. The two frequency distributions of tariff rates (Chart 4.H.1) are fairly similar. They show that in this sector tariffs tended to be rather high – the most common range is 20 to 24.9 per cent. As Table 4.H.1 indicates, nearly all the high rates refer to the products in SITC 732 – Road Motor Vehicles. The distribution of tariff differences (Chart 4.H.2) shows that UK tariffs exceeded EEC rates of duty for the same product in 12 out of the 14 cases, with the +2 and +4 brackets each accounting for four items (most of which are also constituents of the vehicle group).

As usual the UK's and the EEC's export performances are first considered in the aggregate. As Table 4.H.2 demonstrates, the UK had quite a substantial crude trade surplus vis-à-vis the EEC in transport equipment, with exports to the EEC at just over $250 million and trade in the other direction at just over $180 million. Exports to third countries averaged about $1,400 million in the case of the UK and nearly $2,400 million for the EEC. The overall neutral market export ratio (58.6) is higher than that for any of the other industry sections. Both aggregate indicators therefore suggest that the industry as a whole should enjoy a relatively strong position in an enlarged EEC. The ratios shown in column 6 are in a continually descending order and as such indicate good agreement between the two performance tests. At the same time, there is a large gap between the ratios for classes I and II on the one hand and those for classes III and V on the other (according to the mutual trade rank order none of the items is assigned to class IV). This means that the difference between the third country export performance of the items which the mutual trade ranking places in classes I and II and that of the products in classes III and V is unusually large.

Discussion of Table 4.H.3 (which sets out the UK's and the EEC's export performances in detail) might best begin with consideration of two general points. (a) The overall position of the industry as measured by the number of items in each class is clearly in line with the favourable impression conveyed by the aggregate indicators in Table 4.H.2 – classes I and II between them account for over half the items, and the absence of any entry in class IV is accompanied by the assignment of only two products to class V. (b) It so happens that the quantitative significance of the items included in the present section varies to a very marked degree – far more than is found in the other industry sections. The figures for intra-trade which has been taken as a rough quantitative yardstick are less than $1 million for one product and barely $10 million for a second (both refer to chassis with engines) whilst on the other hand there are Vehicle Parts and Passenger Cars with intra-trade in excess of $600 million and $1,100 million respectively. These differences are an unfortunate feature, in the sense that it would have been far better for the purposes of the present study if a breakdown of the large items had been available.

More detailed consideration of the data in Table 4.H.3 shows that of the class I items much the most important are Vehicle Parts and Aircraft, and that class I also contains one conflicting evidence case, trailers (which on the third country export record would move to class IV). With regard to Vehicle Parts it may be noted that the result obtained parallels the assignment to class I of Electrical Equipment for Vehicles (see Table 4.G.3 in the last section) and of Other Internal Combustion Engines (see Table 4.F.3 in the last section but one) of which vehicle engines form an important constituent. These facts suggest that the outlook for British producers of vehicle components as a whole is relatively favourable, though it must be remembered that the second of the items mentioned (Electrical Equipment) is a conflicting evidence case; moreover, a fourth component product – Rubber Tyres – only made class III (see Table 4.B.3).

In the case of Aircraft (also assigned to class I) the interpretation of our findings appears particularly difficult. Large scale government participation in the development of aeroplanes, especially of those designed for commercial and military purposes, together with state ownership of the main commercial airlines, means that the market for Aircraft possesses a number of peculiar characteristics. In consequence it is questionable whether opportunities for British manufacturers will be as favourable as the assignment of their product to class I might be supposed to indicate.

The most noteworthy entries to class II are the two items referring to commercial vehicles other than buses – Lorries and Trucks, and Special Purpose Lorries. A relatively high placing in the rank order for both products is fully confirmed by the neutral market export record, especially in the case of ordinary lorries.

Class III is distinguished by the agreed assignment to it of Passenger Cars (quantitatively the most important of all 230 products) – a result which is perhaps a little surprising in view of the relatively strong position enjoyed by most of the component manufacturers. Class III also contains one conflicting evidence product, Bicycles, which the alternative rank order would place in class I.

The two entries in class V are Motorcycles, and Ships and Boats. The first of these is very much a conflicting evidence case, as the neutral market export ranking would place Motorcycles right at the other end of the scale in class I. With regard to Ships and Boats it must be remembered that for much the greater part of the industry – that producing sea-going vessels – free trade conditions in one sense already existed because neither the EEC nor the UK levied a tariff on sea-going ships.[18] At the same time, other devices (favourable credit terms, subsidies, etc.) have been employed by the authorities both in the UK and in the EEC countries to afford the producers of sea-going vessels some protection. If these measures merely served to act as a substitute for tariffs, no special problems would arise. However, there is some reason to think that at least some of them give particular encouragement to exports. In so far as this is true the use of foreign trade statistics to indicate the comparative cost position of the industry is open to serious reservations, and the pessimistic conclusion normally drawn from the low place of the industry in the rank order may not be valid.

The final topic is trade diversion. Table 4.H.4 sets out, in the upper part of the table, EEC imports from third countries of the items placed in classes I and II (and also of one conflicting evidence product of class III), and in the lower part, UK third country imports of the products in class V, together with the relevant rates of the CET (post-Kennedy Round). The size of EEC third country imports of Aircraft (the second largest of any of the 230 products) at first sight suggests that UK manufacturers face excellent prospects for raising sales at the expense of outside suppliers, but the peculiar nature of the market for aeroplanes which was referred to above casts serious doubt on this conclusion. Vehicle Parts therefore emerge as the only product in the upper part of the table which *prima facie* affords UK exporters scope to divert sales in their favour by substantial margins.

Opportunities open to EEC producers (referred to in the lower part of the table) appear very limited, as the UK's third country imports of the two items shown were well below the average for all goods. Moreover, as one of them (Motorcycles) is a conflicting evidence case, it is not clear that EEC suppliers enjoy a sufficient competitive advantage relative to UK producers.

18 The tariff shown against the item in Table 4.H.1 is thus an average between the zero rate for sea-going vessels and a positive rate for other kinds of boats.

Table 4.H.1 *Transport Equipment: Products Covered and Tariff Rates*

SITC	Description	Tariffs (1965/67; per cent) UK	EEC	UK-EEC
731	Railway vehicles	18.0	12.1	5.9
732.1	Passenger motor cars, whether or not assembled	24.6	23.5	1.1
732.2	Buses, whether or not assembled	24.5	23.5	1.0
732.3	Lorries and trucks, whether or not assembled	24.5	21.0	3.5
732.4	Special purpose lorries, whether or not assembled	24.0	20.0	4.0
732.5	Road tractors for tractor-trailer combinations	17.3	19.0	–1.7
732.6	Chassis with engines mounted of a kind used for vehicles of heading 732.1	25.2	22.0	3.2
732.7	Other chassis with engines mounted	16.3	22.0	–5.7
732.8	Bodies, chassis, frames and other parts of motor vehicles other than motorcycles ('Vehicle parts')	20.0	19.0	1.0
732.9	Motorcycles, motorised cycles and their parts	25.4	20.0	5.4
733.1	Bicycles and other cycles, not motorised	20.0	16.5	3.5
733.3	Trailers and other vehicles, not motorised	23.0	14.4	8.6
734	Aircraft	16.7	14.4	2.3
735	Ships and boats	10.0	2.7	7.3

Notes: See Table 4.A.1.

Chart 4.H.1 *UK and EEC Tariff Rates on Transport Equipment* (1965/67)

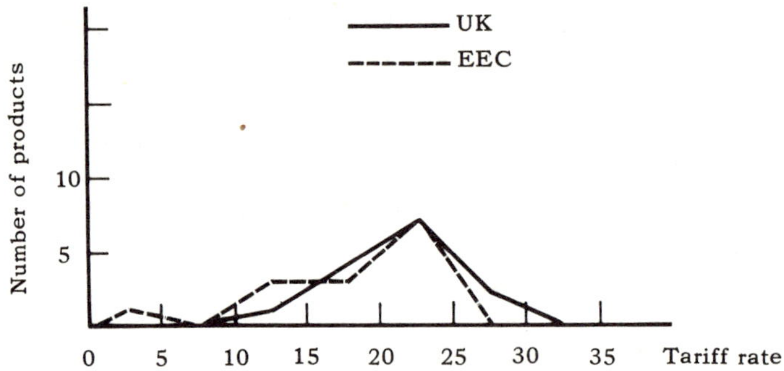

Source: Table 4.H.1.

Chart 4.H.2 *Differences between Tariffs on Transport Equipment* (UK minus EEC, 1965/67)

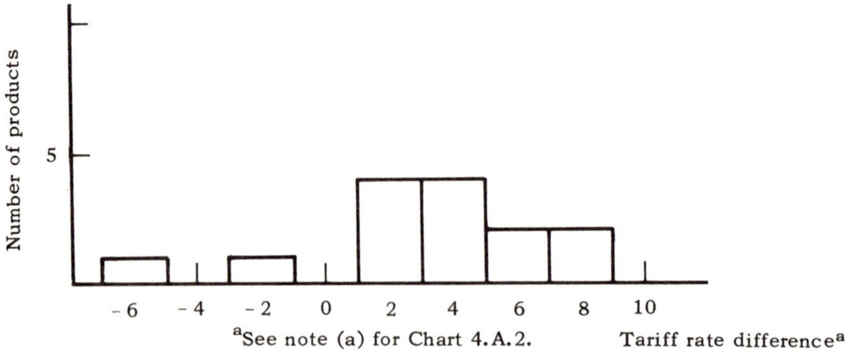

[a]See note (a) for Chart 4.A.2.

Source: Table 4.H.1.

Table 4.H.2 *Trade in Transport Equipment by Class (Mutual Trade Ranking)* (1965/67 average)

Class	No. of items	UK exports to EEC ($million) (1)	EEC exports to UK ($million) (2)	Intra-trade ($million) (3)	UK exports to third countries ($million) (4)	EEC exports to third countries ($million) (5)	(4)/(5) (per cent) (6)
I	4	159.4	69.9	816.8	527.0	509.6	103.4
II	4	11.9	5.9	215.2	311.0	313.5	99.2
III	4	66.9	71.2	1,186.9	444.6	1,229.7	36.2
IV	0	0	0	0	0	0	–
V	2	13.5	34.5	161.1	119.3	337.8	35.3
Total	14	251.7	181.6	2,380.0	1,401.8	2,390.5	58.6
Transport equipment as % of all manufactures		12.3	9.5	12.5	21.5	17.5	–

Notes: See notes (i) to (v) of Table 4.A.3.

Table 4.H.3 *Trade in Transport Equipment: A Detailed Comparison*
(1965/67 average)

Class	SITC	Description	UK exports to EEC ($'000) (1)	EEC exports to UK ($'000) (2)	Intra-trade ($'000) (3)	UK exports to third countries ($'000) (4)	EEC exports to third countries ($'000) (5)	Special features (6)
I	732.6	Car chassis with engines	69	5	830	1,084	1,615	
	732.8	Vehicle parts	116,575	50,845	619,095	278,865	296,344	
	733.3	Trailers	12,093	934	49,605	8,761	26,474	IV
	734	Aircraft	30,659	18,165	147,268	238,253	185,141	
II	731	Railway vehicles	1,539	913	29,074	40,626	90,631	
	732.3	Lorries & trucks	9,072	4,184	161,254	190,449	180,795	
	732.4	Special purpose lorries	947	666	14,585	11,267	22,072	
	732.7	Other chassis with engines	355	139	10,310	68,673	20,013	
III	732.1	Passenger motor cars	64,850	68,905	1,128,755	399,637	1,150,361	
	732.2	Buses	53	37	14,412	12,379	33,302	
	732.5	Road tractors	656	738	18,973	4,748	19,687	
	733.1	Bicycles	1,331	1,510	24,786	27,789	26,336	I
IV	NONE							
V	732.9	Motorcycles	882	7,355	34,923	34,950	47,406	I
	735	Ships & boats	12,602	27,193	126,136	84,368	290,346	

Notes: See notes to Table 4.A.3.

Table 4.H.4 *Transport Equipment: Determinants of Trade Diversion*

(a) In favour of UK sources of supply

Class	SITC	Description	EEC imports from third countries (1965/67 average; $'000)	Post-Kennedy CET (per cent)
I	732.6	Car chassis with engines	85	22.0
	732.8	Vehicle parts	56,144	12.3
	733.3	*Trailers*	2,550	6.5
	734	Aircraft	264,517	7.8
II	731	Railway vehicles	2,317	5.8
	732.3	Lorries & trucks	5,716	10.6
	732.4	Special purpose lorries	3,906	10.0
	732.7	Other chassis with engines	129	11.0
III	733.1	*Bicycles*	3,585	12.5

(b) In favour of EEC sources of supply

Class	SITC	Description	UK imports from third countries (1965/67 average; $'000)	Post-Kennedy CET (per cent)
V	732.9	*Motorcycles*	6,205	10.0
	735	Ships & boats	14,139	2.4

Notes: See notes to Table 4.A.4.

I. Miscellaneous Manufactures

As its heading indicates, the last of the industry sections deals with rather a rag bag of commodities, ranging from cameras and medical instruments to toys and stationery. The total number of products included is 20: detailed descriptions of the items and the appropriate UK and EEC tariff rates are set out in Table 4.I.1.

In view of the miscellaneous nature of the goods covered it is perhaps not surprising that the two frequency distributions of tariffs (Chart 4.I.1) extend over a wide range of rates. This is particularly true of the distribution of the British set of tariffs, which includes some rates near or in excess of 30 per cent (the items concerned belong to the optical goods, photographic equipment, and scientific instruments categories). Differences between UK and EEC tariffs for the same good accordingly also varied a great deal, as Chart 4.I.2 shows. The extreme cases (differences of 15 percentage points or more) with one exception refer to the items already singled out because of the high absolute level of the British tariff.

UK and EEC exports of the products included, *in toto* and by class, are shown in Table 4.I.2. The crude trade balance was just in the UK's favour, with British exports to the EEC averaging $110 million and EEC exports to the UK $101 million. Exports to neutral markets came to $266 million for the UK and $624 million for the EEC (the lowest figures for any of the nine industry groups), and the neutral market export ratio works out at 42.6. This indicates a poorish overall performance for the sector, whereas the crude trade balance gives a somewhat more favourable impression. The order of the ratios in column 6 is somewhat erratic which suggests that agreement between the two performance measures is not very good.

Details of the composition of the five performance classes are shown in Table 4.I.3. Discussion of the table might best begin with a general point. The distribution of the 20 items over the five classes is very uneven; the two extreme classes between them account for nearly three-quarters of all the products, and there is only one entry in each of classes II and III. In other words, the great majority of the British producers of these goods occupy either a relatively very strong or a very weak position. The fact that there are only three conflicting evidence cases, together with the way the ranking of the products concerned is altered, suggests that a similar conclusion would be reached if the classing were determined with reference to third country exports. Broadly speaking this is indeed the case, for the number of products which the alternative rank order would place in each class is as follows.

class I – 5
class II – 3
class III – 1
class IV – 4
class V – 7

Once again class III contains only one entry; on the other hand, the extreme classes, I and V, are slightly smaller than they are in the mutual trade rank order.

Detailed consideration of class I shows that it includes two out of the three entries referring to photographic apparatus, namely Cameras and Other Photographic Equipment (the third, Cine Cameras, is placed at the other end of the scale in class V). This result may at first sight be rather surprising, even though the first item (Cameras) is a conflicting evidence product, but it must be remembered that both entries include photocopying equipment (which in fact accounts for much the greater part of UK exports to the EEC under these headings).[19]

As already mentioned, class II consists of only one item – Other Instruments, though quantitatively it is quite an important product. No comment appears to be called for with regard to class III. The most important item in class IV – Articles of Plastic – is both a border-line case (i.e., it nearly made class III) and a

19 Readers who may wish to extend the figures shown in the table for UK/EEC trade in SITC 861.4 and 861.6 either to earlier or to later years should be warned that during the period to which the figures in the table refer (1965/67) there appears to have been a change in the nature of photocopying equipment, or a change in the classification, such as to transfer the bulk of UK exports to the EEC from SITC 861.6 to SITC 861.4.

conflicting evidence product; the alternative rank order would assign it to class II.

The first point to note about class V is the agreed assignment to it of all three optical goods; it thus appears that British producers of these items (who, as noted earlier, at present enjoy a particularly high level of protection) will be in a relatively weak position when trade barriers are abolished. The makers of clocks and watches similarly look vulnerable. Mention should also be made of the conflicting evidence case, Other Printed Matter, which on the basis of third country exports would move right up the scale to class I. In any case, there must be some doubt how far trade in this item reflects comparative cost conditions.

Data used to examine the prospects for exports being raised as a result of trade diversion are brought together in Table 4.I.4. The upper part of the table (which shows for the goods in classes I and II the EEC's third country imports and the appropriate post-Kennedy Round CET) indicates that there are two items, Other Instruments and Toys, which *prima facie* offer the relevant UK exporters opportunities for increasing exports at the expense of suppliers in third countries by large margins – the EEC's third country imports as recorded are quite sizeable, and the CET is by no means low, especially that on Toys. According to the lower half of the table the UK's third country imports of the goods in classes IV and V were generally fairly low, with the result that the EEC's exporters of any single item appear to enjoy only limited scope for raising sales by means of trade diversion. In combination, however, the opportunities clearly add up to a sizeable total.

Chart 4.I.1 *UK and EEC Tariff Rates on Miscellaneous Manufactures* (1965/67)

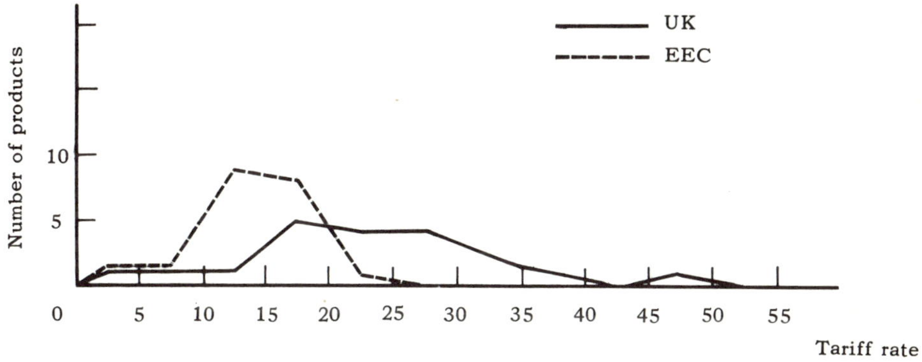

Source: Table 4.I.1.

Table 4.1.1 *Miscellaneous Manufactures: Products Covered and Tariff Rates*

SITC	Description	Tariffs (1965/67; per cent) UK	EEC	UK-EEC
861.1	Optical elements	45.8	16.5	29.3
861.2	Spectacles and spectacle frames	20.0	17.0	3.0
861.3	Binoculars, microscopes, and other optical instruments	37.0	15.4	21.6
861.4	Photographic cameras (other than cinematographic) and flashlight apparatus	24.0	17.0	7.0
861.5	Cinematographic cameras, projectors, sound recorders and sound reproducers	30.5	14.0	16.5
861.6	Other photographic and cinematographic apparatus and equipment	32.5	13.0	19.5
861.7	Medical instruments, n.e.s.	16.5	13.0	3.5
861.9	Measuring, controlling and scientific instruments, n.e.s.	28.8	13.5	15.3
862.3	Chemical products and flashlight materials of a kind used in photography put up in measured portions or for retail sale	15.0	12.0	3.0
862.4	Photographic film, plates and paper, whether or not exposed, and developed film other than developed cinematographic film	12.5	14.5	−2.0
863.0	Developed cinematographic film	0	0	–
864.1	Watches, watch movements and cases	27.5	12.0	15.5
864.2	Clocks, clock movements and parts	27.5	16.0	11.5
891.8	Musical instruments, n.e.s.	25.3	15.0	10.3
892.9	Printed matter, n.e.s. (paper labels, plans and drawings for industrial purposes, unissued postage stamps, calendars, etc.)	7.9	10.0	−2.1
893.0	Articles of artificial plastic materials, n.e.s.	19.7	17.6	2.1
894.2	Children's toys, indoor games, etc.	24.0	20.0	4.0
894.4	Other sporting goods (fishing and hunting equipment and other requisites for outdoor sports)	20.5	16.3	4.2
895.2	Pens, pencils and fountain pens	17.9	12.1	5.8
897.1	Jewellery of gold, silver and platinum group metals and goldsmiths' or silversmiths' wares including set gems	19.9	9.5	10.4

Notes: See Table 4.A.1.

Chart 4.I.2 *Differences between Tariffs on Miscellaneous Manufactures* (UK minus EEC, 1965/67)

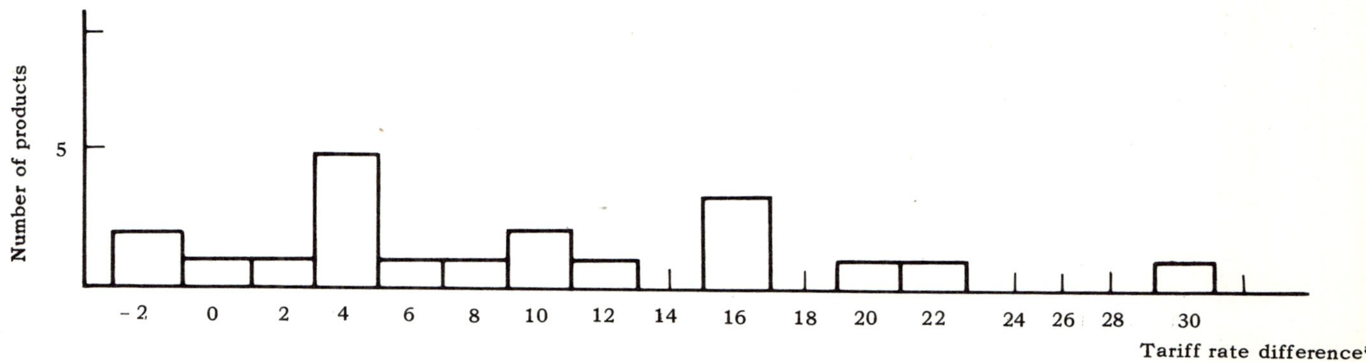

[a] See note (a) for Chart 4.A.2.
Source: Table 4.I.1.

Table 4.I.2 *Trade in Miscellaneous Manufactures by Class (Mutual Trade Ranking)* (1965/67 average)

Class	No. of items	UK exports to EEC ($million) (1)	EEC exports to UK ($million) (2)	Intra-trade ($million) (3)	UK exports to third countries ($million) (4)	EEC exports to third countries ($million) (5)	(4)/(5) (per cent) (6)
I	6	52.1	18.6	196.2	68.5	110.1	62.3
II	1	26.4	16.4	142.0	79.8	92.3	86.4
III	1	2.7	3.1	18.9	11.7	31.3	37.3
IV	4	16.8	31.1	323.0	53.3	214.0	24.9
V	8	12.0	31.5	193.6	52.4	175.9	29.8
Total	20	110.0	100.8	873.6	265.7	623.6	42.6
Miscellaneous manufactures as % of all manufactures		5.4	5.4	4.6	4.1	4.6	–

Notes: See notes (i) to (v) of Table 4.A.3.

Table 4.I.3 *Trade in Miscellaneous Manufactures: A Detailed Comparison* (1965/67 average)

Class	SITC	Description	UK exports to EEC ($'000) (1)	EEC exports to UK ($'000) (2)	Intra-trade ($'000) (3)	UK exports to third countries ($'000) (4)	EEC exports to third countries ($'000) (5)	Special features (6)
I	861.4	Cameras, etc.	17,002	4,583	44,097	9,094	27,404	IV
	861.6	Other photo equipment	12,422	5,522	37,010	7,869	16,976	
	862.3	Chemicals for photography	1,160	566	6,602	2,183	2,871	
	863.0	Developed film	5,456	1,723	16,152	8,917	9,382	
	894.2	Toys	13,826	5,054	78,493	28,335	34,983	
	894.4	Other sporting goods	2,203	1,172	13,826	12,137	18,465	*
II	861.9	Other instruments	26,406	16,439	141,975	79,771	92,282	
III	861.7	Medical instruments	2,654	3,133	18,866	11,676	31,315	
IV	862.4	Photo film, etc.	7,681	15,152	111,775	28,425	88,116	
	893.0	Articles of plastic	7,232	10,015	138,017	15,620	35,045	† II
	895.2	Pens & pencils	1,515	2,317	21,778	4,076	25,645	
	897.1	Jewellery	415	3,578	51,384	5,174	65,201	
V	861.1	Optical elements	1,628	2,216	14,014	3,997	11,948	†
	861.2	Spectacles	285	3,392	18,165	630	29,278	
	861.3	Optical instruments	1,191	3,107	19,417	2,889	22,945	
	861.5	Cine cameras, etc.	1,086	3,217	15,123	1,626	12,199	
	864.1	Watches	416	1,298	12,744	2,129	20,463	
	864.2	Clocks	3,412	5,805	42,151	8,795	36,772	†
	891.8	Musical instruments	513	1,772	12,590	2,597	22,794	
	892.9	Other printed matter	3,477	10,694	59,409	29,755	19,472	I

Notes: See notes to Table 4.A.3.

Table 4.I.4 *Miscellaneous Manufactures: Determinants of Trade Diversion*

(a) In favour of UK sources of supply

Class	SITC	Description	EEC imports from third countries (1965/67 average; $'000)	Post-Kennedy CET (per cent)
I	861.4	*Cameras, etc.*	18,578	10.5
	861.6	Other photo equipment	14,334	8.8
	862.3	Chemicals for photography	3,098	9.6
	863.0	Developed film	4,446	n.a.
	894.2	Toys	52,764	12.7
	894.4	Other sporting goods	10,964	8.1
II	861.9	Other instruments	84,872	8.7

(b) In favour of EEC sources of supply

Class	SITC	Description	UK imports from third countries (1965/67 average; $'000)	Post-Kennedy CET (per cent)
IV	862.4	Photo film, etc.	13,206	9.4
	893.0	*Articles of plastic*	10,806	13.6
	895.2	Pens & pencils	4,201	6.7
	897.1	Jewellery	3,311	6.0
V	861.1	Optical elements	4,033	12.8
	861.2	Spectacles	1,327	8.5
	861.3	Optical instruments	5,280	11.2
	861.5	Cine cameras, etc.	8,076	10.5
	864.1	Watches	2,572	8.4
	864.2	Clocks	3,865	9.2
	891.8	Musical instruments	2,230	7.9
	892.9	*Other printed matter*	5,978	6.0

Notes: See notes to Table 4.A.4.

5 Conclusion

The final chapter of the study is devoted to three aims – (a) to introduce a set of summary tables, (b) to compare the results of the study with other indicators of the likely positions of different British industries in the event of British accession to the EEC, and (c) to mention a few of the ways along which further research into the problem might fruitfully proceed.

(a) *Summary tables*

One straightforward way of comparing the relative positions of the nine industry sectors in an enlarged EEC is to calculate, by number of items, the proportion of each industry's products which is assigned to the various performance classes. The outcome is shown in Table 5.1.

Table 5.1 *Class Distribution of Products by Number*
(Mutual Trade Ranking)

per cent

Industry	No. of products	Class I	II	III	IV	V	Total
Chemicals	38	23.7	10.5	15.8	26.3	23.7	100.0
Manufactures of leather, rubber, wood, paper, etc.	32	21.9	18.8	15.6	15.6	28.1	100.0
Textiles and clothing	30	13.3	23.3	30.0	16.7	16.7	100.0
Iron and steel	24	4.2	41.7	37.5	4.2	12.6	100.0
Non-ferrous metals and metal manufactures	24	33.3	29.2	16.7	16.7	4.2	100.0
Mechanical engineering	29	10.3	17.2	10.3	41.4	20.6	100.0
Electrical engineering	19	21.1	10.5	26.3	26.3	15.8	100.0
Transport equipment	14	28.6	28.6	28.6	0	14.3	100.0
Miscellaneous manufactures	20	30.0	5.0	5.0	20.0	40.0	100.0
Total	230	20.0	20.0	20.0	20.0	20.0	100.0

Source: Tables 4.A.3, 4.B.3, 4.C.3, etc.

Table 5.2 *Class Distribution of Products by Value*
(Mutual Trade Ranking)

per cent

Industry	As % of total intra-trade in manufactures	Class I	II	III	IV	V	Total
Chemicals	11.6	16.1	11.4	12.3	35.6	24.7	100.0
Manufactures of leather, rubber, wood, paper, etc.	10.3	32.0	26.0	14.8	7.4	19.9	100.0
Textiles and clothing	13.2	13.3	12.5	41.7	19.2	13.3	100.0
Iron and steel	10.1	0.7	36.6	39.5	18.9	4.3	100.0
Non-ferrous metals and metal manufactures	8.9	27.1	31.0	26.5	14.8	0.6	100.0
Mechanical engineering	19.3	14.7	22.3	9.0	37.0	16.9	100.0
Electrical engineering	9.5	13.3	15.2	44.6	14.9	12.1	100.0
Transport equipment	12.5	34.3	9.0	49.9	0	6.8	100.0
Miscellaneous manufactures	4.6	22.5	16.3	21.6	37.0	22.2	100.0
Total	100.0	19.5	17.9	29.2	20.3	13.0	100.0

Source: Tables 4.A.3, 4.B.3, 4.C.3, etc.

Before the table is considered attention should be drawn to two defects of the procedure followed – (i) the information presented treats all products alike, despite the fact that large differences exist in the quantitative significance of the individual goods, and (ii) the table is wholly based on the mutual trade rank order and no account is taken of the information supplied by relative UK and EEC exports to third countries. The finding that neutral market export performances seriously challenge some of the evidence supplied by the mutual trade data, notably in the case of the iron and steel industry, is thus disregarded.

The first of these difficulties has been tackled by also comparing the nine industry groups in terms of the proportion of the value of total trade (as hitherto defined as UK-cum-EEC intra-trade) for which the items in each class are responsible. The result is shown in Table 5.2.[1] The second problem cannot be satisfactorily dealt with by drawing special attention to conflicting evidence products, as this would make the table cumbersome and difficult to comprehend. Initially we therefore considered the possibility of inserting two further tables based on the same criteria as Tables 5.1 and 5.2 but with the classing wholly determined by relative export performances in neutral markets. We ultimately rejected the idea because it would mean the introduction at this late stage of the work of a largely

1 Some readers may wonder why Table 5.1 was included at all, in view of the objections to treating all products alike. The answer is largely a negative one. The use of intra-trade statistics as indicators of quantitative significance is itself open to serious misgivings, as intra-trade is likely to be a very imperfect proxy of the weights one would really want to employ – the value of domestic production and/or consumption; *ipso facto* the case for simply considering the number of items in each class is strengthened.

new body of material.[2] In the end the problem posed by conflicting evidence cases has therefore been left unresolved.

There is no need for Tables 5.1 and 5.2 to be discussed in any detail; the figures can be left to speak for themselves, and in any case much of the material is already familiar. This applies particularly to Table 5.1 which shows class distributions by number of products, and comment will therefore largely be confined to Table 5.2.

It is evident that the table supports the main conclusion to be drawn from Chapter 4 – whilst some of the nine industry groups are rather more favourably placed vis-à-vis EEC manufacturing than are others, the position of Britain's manufacturing industries relative to the EEC's is not a matter of right-across-the-board strength of some sectors accompanied by marked weakness of others, but rather of each sector having its more promising and its relatively vulnerable parts (at least in terms of the industrial classification adopted). As to the identity of the sectors with relatively large numbers of well and lowly-placed constituents respectively, the picture presented by Table 5.2 differs somewhat from that conveyed by Table 5.1 – which in view of the variations in the quantitative significance of the 230 items is hardly surprising.

Nevertheless, the application of a set of admittedly rather crude criteria reveals certain broad patterns. If for each industry group the products assigned to classes I and II are taken together (i.e., differences between classes I and II are disregarded), the three best performers according to Table 5.1 are Iron and Steel, Non-Ferrous Metals and Metal Manufactures, and Transport Equipment. A similar list emerges if these criteria are applied to Table 5.2, the difference being that Iron and Steel is replaced by Manufactures of Leather etc. At the other end of the scale agreement between the two tables is even closer – the three sectors with the highest percentage of items in classes IV and V, by number or by value of intra-trade, are Chemicals, Mechanical Engineering, and Miscellaneous Manufactures. Textiles and Clothing, and Electrical Engineering, are thus left as the two sectors which emerge as middling performers.

All these conclusions are subject to the important proviso that the data take no account of the evidence supplied by relative exports to third countries. In the case of Iron and Steel this objection is so strong (see Section D of Chapter 4) that the industry should perhaps have been omitted from the tables, but considerable caution is also called for with regard to other industry groups. For instance, as Section F of Chapter 4 shows, Mechanical Engineering products in general perform rather better in the third country export comparison than they do in the mutual UK/EEC test; in consequence, if Tables 5.1 and 5.2 had been drawn up on the basis of the neutral market rank order, the industry would be shown to be one of the three strongest performers, rather than being among the three worst (see Appendix Tables B.4 and B.5).

The finding that each sector has its strong and its weak parts calls for one further comment. It has repeatedly been shown that so far the main effect of the

2 Two tables drawn up along the lines indicated are, however, included in the Appendix (Tables B.4 and B.5).

creation of the EEC upon the economies of the six member countries has been redistribution of resources within industries rather than between industries – *intra*-industry rather than *inter*-industry specialization.[3] The predictions of the present study regarding manufacturing industry in Britain and in the EEC as a whole in the event of Britain's entry are thus fully consistent with the EEC's past experience.

Moreover, the data suggest that there would be a tendency for specialization to take place at the level of relatively specific products rather than of broader product groups – in the language of the SITC, at the 4-digit rather than at the 3-digit level (this again is in line with past experience). An examination of the detailed performance tables in Chapter 4 indicates that there are only eleven 3-digit groups (out of 52) all of whose constituents are uncontestedly assigned to one given class.[4] If slightly less stringent criteria are applied and the agreed placing in adjoining classes is taken as the basis, the number of SITC groups whose constituents are alike in the sense described rises to 17 – still less than one-third of the total.[5] This finding has important implications for the adjustment process which would follow the dismantling of tariff barriers between the UK and the EEC. Unfortunately it also places severe practical limitations on the possibility of exploring the effects of the predicted pattern of specialization, a matter considered under (c) below.

(b) *Comparisons of results*

The second topic for discussion in this chapter is the comparison of the results obtained with other evidence relating to the effect of British accession to the EEC upon manufacturing industry in the UK and in the Common Market. As implied in Chapter 1, it does not appear that the subject matter of the present study has been investigated in detail by other research workers, and reference to the earlier work cited in Chapter 1 (cf pp. 7–8) would be of only limited value because of the time which has elapsed since those researches were carried out, quite apart from the somewhat different terms of reference. However, shortly before the study went to press an article appeared in the Board of Trade *Journal* (henceforth referred to as BTJ) which in the course of a general discussion of the EEC's trade referred to a number of products, British exports of which would be likely to do particularly

3 See B. Balassa, 'Tariff Reductions and Trade in Manufactures among the Industrial Countries', *American Economic Review*, June 1966, pp. 466–473; H. G. Grubel 'Intra-Industry Specialization and the Pattern of Trade', *Canadian Journal of Economic and Political Science*, August 1967, pp. 374–388; M. Adler, 'Specialization in the European Coal and Steel Community', *Journal of Common Market Studies*, March 1970, pp. 175–191; and I. Walter, *The European Common Market: Growth and Patterns of Trade and Production*, New York, 1967, Chapter 6.

4 SITC groups whose representation is limited to one sub-group are omitted.

5 The possibility that specialization would in part at least occur at an even more disaggregated stage is, of course, not precluded.

well in the event of the membership negotiations succeeding.[6] Although it may not be appropriate to press the information in BTJ very far (particularly as it is not clear how the classification was arrived at), there appears to be some merit in comparing it with the relevant results obtained in the present study (briefly referred to as DAE), if only in order to put the DAE findings into sharper focus.[7]

The Board of Trade confined themselves to products in which they thought the UK would likely to be performing particularly well. The easiest way of tackling the comparison is therefore to set out in tabular form the products listed in BTJ and place against each item the result obtained in DAE.

Table 5.3 *Industrial Effects of EEC Entry: A Comparison*

BTJ Description (1)	SITC (2)	DAE Class (3)	Remarks (4)
China & earthenware (B)	666.4	(III)	tcX: I; DAE only refers to china
Footwear (B)	851.0	V	
Ready-to-wear clothing (C)	841.1	III	
Hosiery (C)	841.2	(IV)	DAE refers to all clothing accessories
Knitwear (C)	841.4	III	tcX: V
Computers & peripheral equipment (F)	714.2	(IV)	DAE: 'calculating machines'
	714.3	(V)	DAE: 'statistical machines'
Machine tools (F)	715.1	V	tcX: III
Textile machinery (F)	717.1	IV	tcX: II
Printing equipment (F)	718.2	(IV)	DAE also refers to bookbinding equipment
Food processing equipment (F)	718.3	V	tcX: III
Construction machinery (F)	718.4	(III)	DAE also refers to mining machinery
Goods handling equipment (F)	719.3	II	
Bottling & packaging equipment (F)	719.6	(IV)	DAE: part of 'other non-electrical machines'
Automated office equipment (F, I)	714.9	(III)	tcX: I; DAE: 'other office machinery'
	861.4	(I)	tcX: IV; DAE: 'cameras'
	861.6	(I)	DAE: 'other photographic equipment'
Scientific instruments (I)	861.3	(V)	DAE: 'optical instruments'
	861.7	(III)	DAE: 'medical instruments'
	861.9	(II)	DAE: 'other instruments'
Toys (I)	894.2	I	
Sports equipment (I)	894.4	(I)	DAE: 'other sporting goods'

Sources and methods: See Ch.5, pp. 100–103.

6 See Board of Trade *Journal*, 8 July 1970, pp. 41–43.

7 Attempts to draw comparisons with another set of evidence concerning the industrial effects of Common Market entry (see Institute of Directors, *The Balance Sheet of the Common Market*, London, December 1969) proved abortive, largely because of differences in the classification of commodities.

This has been done in Table 5.3. Column 1 shows the commodities or commodity groups singled out in BTJ as promising (the capital initial placed in brackets after each item refers to the relevant section of DAE Chapter 4), column 2 identifies the product by its SITC number,[8] column 3 states the class to which the item is assigned in the DAE mutual trade rank order, and column 4 adds any qualifying remarks. These are of two kinds. To begin with, any conflicting evidence items are identified by the initials tcX (short for third country exports), with the alternative class in Roman numerals. Thus, tcX III means that the third country export ranking would place the item concerned in class III. Secondly, differences in coverage are indicated by notes stating in what way the description of the corresponding items in DAE diverge. In these cases the two sets of information cannot really be compared, and the classes in column 3 have therefore been bracketed.

The first point to be made with reference to Table 5.3 is that it only covers a small fraction (less than 10 per cent by number and less than 15 per cent by value of intra-trade) of the products included in the present study. Moreover, the sample is confined to constituents of four industry sections, and the other five are not represented.[9]

The extent of agreement between the BTJ and DAE indicators is clearly limited. Column 3 shows that if bracketed items are excluded there are only two products which DAE would unambiguously place in classes I or II (which may be taken as equivalent to the BTJ description of a product as 'particularly likely to surge forward') whilst five of the remaining six products at best make class III (the sixth, Textile Machinery, is assigned to class IV by the mutual trade rank order and to class II by the third country export test). The nine bracketed items do not lend themselves to a straightforward summary, but the information set out is certainly not such that one could speak of broadly similar conclusions emerging from the two sources.

The upshot is therefore that in the case of the small sample covered there are substantial divergencies between the BTJ predictions concerning the effects of EEC entry and those emerging from the present study. There are no doubt many reasons which could be responsible for this; one of them deserves to be singled out. The BTJ description of the items mentioned in the article as 'particularly likely to surge forward' clearly refers to the direction of the EEC effect, whereas the DAE rank order relates to past performance and therefore (at best) to the relative competitive position of the 230 items; as repeatedly pointed out in Chapter 3, the link between the placing of a product in the rank order and the direction of the change in output after entry is not a very firm one. The next step in the

8 When the BTJ and the DAE classifications differ the SITC number refers to the DAE definition of the product.

9 BTJ also refers to 'electronic equipment' and 'pleasure boats and marine equipment', but these items could not be included in the table; the description 'electronic equipment' is too vague to allow reasonably accurate identification of the products concerned in the SITC classification, and the result for pleasure boats is likely to be swamped by the weight of the remainder of the appropriate SITC entry – shipbuilding. (Marine equipment may, moreover, be part of another SITC group).

comparison between BTJ and DAE would therefore be to explore whether any special circumstances apply in the markets for the products selected in BTJ.[10] More generally, the lack of agreement between BTJ and DAE points to the urgency of further investigation of the problem tackled in this study; the concluding section of the chapter *inter alia* suggests some of the ways along which such work might suitably proceed.

(c) *Further research*

Two issues may suitably be tackled under the heading 'further research' – (i) extensions of the analysis of comparative foreign trade performances and (ii) investigation of the consequences of the structural effects of EEC entry.[11]

(i) It is clear that the method of using and interpreting foreign trade statistics which has been employed in the present study could be modified and extended in a number of ways. In the first place, the existing data could be subjected to rather more careful scrutiny. For instance, the precise coverage of an SITC heading may not be apparent even if the full SITC list is consulted, and one should therefore turn to supplementary sources of information, such as the *Annual Statement of the Trade of the United Kingdom* and the EEC's detailed trade statistics (referred to below). Again, conflicting evidence cases could be considered further; study of the composition of the item, in the manner just indicated, and of principal markets/sources of supply is likely to resolve at least some of the uncertainty.

Secondly, the ranking of the 230 products on the basis of average trade performances for the period 1965/67 could be supplemented by data relating to changes in trade flows over time, as a sufficiently long run of comparable statistics should now be available. At least one major difficulty would arise, however – the combination of the two sources of information into one set of indicators.[12]

10 With regard to trade diversion opportunities it is easy enough to check whether EEC imports of goods with regard to which BTJ and DAE disagree have been relatively large, thus affording UK exporters rather better scope for raising exports at the expense of third country suppliers than would be true of the average product. (The argument that comparative costs play a part in determining how far a set of producers can exploit trade diversion opportunities is naturally disregarded.) The test was not very conclusive – there was some tendency for EEC third country imports of the goods placed differently by BTJ and DAE to lie above the average for all products, but this was not very marked. Relatively favourable trade diversion opportunities are therefore not likely to be the only factor responsible for the divergence between BTJ and DAE.

11 At least one other topic could have been raised at this point – the possibility of investigating the industrial effects of UK accession to the EEC by means other than analysis of foreign trade data. However, alternative approaches and some of their difficulties have already been touched upon in Chapter 1 (see pp. 6–8); there is thus no need at the present stage to go into the question. On the other hand, the various problems and uncertainties encountered in the course of the present study clearly serve to emphasize the need for the exploration of the other methods; at a minimum the results obtained above should be compared with the predictions of the industrialists themselves, both in the UK and in the EEC.

12 For an entirely arbitrary solution to the problem see H.H. Liesner, 'The European Common Market and British Industry', op. cit.

Thirdly, during the last few years a new set of trade statistics issued by the EEC Commission has become available, with a breakdown of commodities which is much more detailed than that used in SITC–based data, and which moreover is at least in principle readily comparable with the Brussels tariff nomenclature. As the statistics only cover the EEC's trade, analysis would have to be confined to the EEC's imports from/exports to the UK, i.e., to mutual trade, but with over 3,000 manufactures separately identified the gain in the comparability of trade flows would be most valuable. Moreover, the achievement of a further level of disaggregation would permit more conclusive analysis of the problem of *inter* versus *intra*-industry specialization (see pp. 99–100 above). Presentation of the results would constitute rather an awkward problem, in view of the sheer volume of figures, but this might be partially overcome by the grouping of products into a large number of industries – far larger than the nine used on the present occasion.

The present may also be a convenient point in the argument to indicate how an industrial user of this study – say, an economist employed in the research department of a British firm – may set about answering with its help the question how UK accession to the EEC is likely to affect the firm concerned.[13] To begin with he would clearly have to identify his firm's products among the 230 goods covered. Typically there would be some overlapping of categories, and in order to resolve the problem satisfactorily he would be well advised to go back to the full SITC (see Ch.3, fn.19), but it will be assumed below that this difficulty has been overcome – say, the firm's one product, sprockets, happens to correspond to a like-named entry in the trade returns. The second step would be to bring additional information to bear on the question of the relative competitive position of British sprockets, i.e., to check whether the assignment of sprockets to, say, class II is about right. In other words, he has to ask himself whether the foreign trade performance of sprockets has been in some sense unusual in one way or the other during the period considered (1965/67).[14] A convenient way of taking up this question would be to compare the place of sprockets in the mutual trade rank order with their position in the third country export rank list. However, it would also seem advisable to bring later data to bear on the issue. For instance, it would be worth while to ask whether the 1967 devaluation – which, on the export side, should presumably have produced results broadly similar to those of a cut in the tariffs which other countries levy on UK goods – affected overseas sales of sprockets

13 *Mutatis mutandis* the paragraphs which follow also apply to a continental user, of course. The argument is restricted to general points; in any particular case there may well be other aspects of at least equal importance.

14 This is not as straightforward as it may appear, because the real question is whether conditions have been *more* unusual than those of other products, and he may not be in a particularly good position to supply an answer for all the other commodities.

in a manner markedly different from that experienced by UK exporters as a whole.[15]

Assume the conclusion is that class II is about right. The next step is to ask how exports to and imports from the EEC will be affected by UK accession. For this purpose he would first of all have to project exports to and imports from the EEC to the late 1970s on a 'no-entry' basis. Subsequently he would have to ask himself how, given the relatively strong competitive position enjoyed by UK sprockets in the market of an enlarged EEC – a position which is assumed not to alter – exports and imports would react to the abolition of sprocket tariffs between the UK and the EEC (the tariffs being removed would be the post-Kennedy Round rates). At each step he would have to bring his knowledge of the EEC and UK markets for sprockets to bear on the problem. For instance, if UK sprocket producers were forecast even without EEC entry to enjoy a substantial share of the EEC market, either at the expense of EEC producers or that of third country suppliers, the projected growth of exports of sprockets as a result of entry would clearly work out at a lower figure than it would if this were not the case. (Difficulties of penetrating a relatively 'unknown' market may, of course, swing the argument the other way.)

However, the exercise could not be regarded as complete without account being taken of at least three further factors. In the first place there is the question of what bearing supply factors will have on the outcome. British firms exporting sprockets to the EEC may find that the necessary rise in output cannot in fact be achieved, at least not very quickly, or not without costs and prices going up. Some examination should thus be undertaken of possible supply difficulties, even though the period of transition to full free trade between the UK and the EEC will presumably extend over several years (the overall balance of the economy should be assumed to be preserved by appropriate government action). A similar assessment is called for with regard to the supply of EEC exports to the UK. The possibility of switching exports between different markets (see Chapter 2, p. 11) should be allowed for.

Secondly, any trade effects calculated in the manner described can only be 'first round' consequences, and the possibility of further changes in exports and imports of sprockets, following upon any adjustment in relative prices to cope with balance of payments disequilibria caused by EEC entry, should be introduced into the model. For instance, it might be thought that UK accession to the EEC would in due course lead to a deterioration of the UK payments position and to an improvement of the EEC's. In that event a rise in the general EEC price level relative to the UK's would be called for; the estimate of the increase in exports of sprockets would therefore have to be further written up, whilst the figure for the rise in imports would be reduced. In the final position British producers of sprockets would thus have a bigger share of the total European sprocket market than they would have done in the absence of the relative downward shift of British prices.[16] The opposite would be the case if overall balance of payments

15 As already indicated (Ch. 3, fn. 7) in general one would not expect the post-devaluation rank order to look markedly different from those obtained in this paper (indeed, if it did a major question mark would be placed against the present study); at the same time such differences as would appear might turn out to provide some useful hints as to the *size* of the additional trade flows generated by EEC entry, at least on the export side.

16 Without a much more careful specification of the situation it is not possible to say how far the enlargement of the UK's share would be at the expense of domestic EEC producers rather than of third country suppliers.

effects were such as to call for a rise in British prices relative to the EEC's.

The size of these 'second-round' changes in trade flows would clearly depend on the extent of the price adjustment, which in turn would be a reflection of the size of the deterioration (or improvement) of the UK's external payments balance as a result of EEC entry. Estimates of this are bound to be uncertain (partly for reasons explained earlier in this study), but probably no more so than many of the other figures the sprocket firm's economist would have to use.

Finally, there is the obvious point that EEC entry may well affect the individual firms making up the UK sprocket industry in rather different ways. The economist must therefore ask himself how his own firm compares with the sprocket industry as a whole in terms of average costs of production, and whether there are any special factors determining the demand for his firm's sprockets, both in the UK and in the EEC.

In general the calculation of the effect which EEC entry would have on trade and output of sprockets is clearly an exceedingly speculative exercise, and the only real argument in favour of proceeding despite the objections is the obvious one that a crude estimate is presumably better than none at all.

(ii) The question of research into the consequences of the industrial effects resulting from the enlargement of the Community opens up an entirely new field for consideration, and discussion must be confined to a few of the more obvious points. One important issue is the relationship between the pattern of specialization and the distribution of resources, in particular of labour and of capital, between different industries. The problem itself is in principle straightforward: falling production (due to demand switching to imports) of certain goods – say, Optical Instruments – will release capital and labour, whilst the increase in the output of, say, Paints (brought about by rising export demand) will require larger capital and labour inputs, but it is unlikely that the proportions in which factors become available are the same as those in which they are used in the growing activities.[17] The question is how far the lines of production which would decline are primarily labour intensive or capital intensive, and vice versa with regard to products which are strongly placed. If expanding activities are capital intensive,

17 The issue is put in static terms as this is much the easiest way of presenting it, but in practice it will often be a matter of rates of growth of output being reduced or being accelerated. The possibility that the industrial consequences of British accession to the EEC primarily take the form of the efficiency in the use of resources in existing activities and at existing scales of production being raised (variously referred to as the import competition or 'cold shower' effect) is being disregarded. Insofar as this effect is important, the points made under (ii) are less significant, but the ranking of products in terms of comparative costs (i.e., the material of Chapter 4) is still useful because it tells one where the pressure towards greater efficiency is most likely to be felt. For an analysis of the postulated relationship between free trade and internal efficiency see W.M. Corden, 'The Efficiency Effects of Trade and Protection', in I.A. McDougall and R.H. Snape *ed.*, *Studies in International Economics*: *Monash Conference Papers*, Amsterdam, 1970, pp. 1–10. As both in the EEC and in EFTA trade among the member countries has risen much faster than trade with third countries, it would appear that, at the very least, the import competition effect does not exhaust the industrial consequences of the abolition of customs barriers.

as is sometimes predicted,[18] any tendency towards labour scarcity in the economy will be reduced, but at the same time capital resources will, relatively speaking, be more stretched, whereas the opposite will be the case if the products on which the economy specializes are labour intensive.

It might be added that in the former case the transition period may present some particularly awkward issues. During that time investment spending is in any event likely to be above its normal rate because of accelerated scrapping of plant in some lines of production and the need for rapid additions to capacity in others; if furthermore the 'new' lines of production are primarily capital intensive the quantity of resources which have to be diverted to investment from other uses (private or public consumption) may be very substantial; this in turn would raise important short-run problems for economic policy makers.

A second important problem relates to scale and technological progress. Economic activities differ with regard to the extent to which unit costs can be expected to fall as output expands, and with regard to the speed with which new processes which allow a given output to be produced with a similar input of resources are found and applied; the question therefore arises in what way the predicted pattern of specialization affects these two determinants of the economy's long-run growth rate.[19]

A third issue is the regional impact of the relative shifts in production. There is no *a priori* reason why expanding activities should counterbalance contracting activities at the regional level; in view of the importance which successive British (and EEC) governments have attached to the attainment of a reasonable degree of regional balance it is clearly of importance to know how different regions would be affected. In particular, the question arises whether there appears to be any tendency for declining activities to be concentrated in parts of the country which in any case find it difficult to attain full employment and a satisfactory growth rate.

Two comments are called for on the points made in the present section. In the first place, it is evident that research into the issues outlined necessitates the matching-up of trade data (given that the comparative cost analysis relies on foreign trade statistics) with industrial statistics. There is little doubt that this will be a difficult undertaking, for industrial statistics as finely broken down as 4-digit SITC export and import figures, let alone the EEC's trade data, are frequently not available even in census of production records. Regrouping and amalgamation of the trade data cannot solve the problem, as the pattern of specialization whose effects are supposedly being assessed will then become obscured. The only solution would appear to be to establish for each trade product group the relevant characteristics (factor intensity, the availability of economies of scale, etc.) of the nearest – generally rather bigger – heading in the industrial

18 See Institute of Directors, *The Balance Sheet of the Common Market*, op. cit., p.9.

19 Economies of scale associated with growing output *per se* – of the type stressed by Kaldor – should in any case be available, simply as a result of the process of specialization. Cf. N. Kaldor, *Causes of the Slow Rate of Economic Growth of the United Kingdom Economy*, Cambridge, 1966.

classification, a messy and in many ways unsatisfactory procedure.

Secondly, it should be made clear that estimation of the effects of British accession to the EEC upon such matters as resource use, opportunities for scale economies, etc. presupposes that one of the central weaknesses of the present study – the failure to tackle the size of trade and output effects – has been overcome. As explained in Chapter 3 (see pp.17–20), in the absence of a solution to this problem it is not possible to predict, except in terms of probability, even the *direction* of the consequences of entry for particular commodity groups. It would thus appear that first priority should be accorded to the derivation of at least crude estimates of the extent to which trade flows would respond to the reduction of trade barriers (possibly by using past EEC and/or EFTA experience as a guide). Subsequently the changes in exports/imports would have to be related to industrial output (which, as just mentioned, raises the question of the availability of appropriate industrial statistics), and from the output changes thus calculated one could then go into questions of resource effects etc. All of which demonstrates the need for a great deal more thought and work to be devoted to the whole topic.

Appendix A: A Note on Tariffs

The main purpose of this annex is to introduce the tariff data used in the study. For reasons stated in Chapter 3 it was necessary to find for each of the 230 commodities the appropriate UK and EEC tariff rates (in the case of the EEC rates ruling in 1965/67 and also the post-Kennedy Round tariffs).[1] In the great majority of cases the breakdown of the tariff data was found to be much more detailed than the commodity trade statistics available. In other words, in general there were several tariff headings for each product, and these had to be reduced to one single rate. In principle several methods may be adopted for calculating an average tariff, none of them fully satisfactory, but in our particular case all except one ruled themselves out because of lack of data, and the only course open to us was the calculation of simple unweighted averages.[2]

It is very likely that the approach adopted has led to a number of odd results, particularly perhaps with regard to some of the semi-manufactures. For all intents and purposes only one of the several rates of duty prescribed for a particular commodity may have any real significance, as the others may refer to versions of the good which are quantitatively of no significance. However, only an expert in the particular branch of industry concerned would be in a position to indicate which tariff rates count and which might be disregarded. Moreover, it should be emphasized that what matters for present purposes is each prospective partner's tariff on the other's goods, and the trade flows which are relevant are therefore those between the UK and the EEC, existing or potential, and not all UK and EEC imports.

For the greater part of the period 1965/67 UK imports of EEC manufactures were impeded not only by the British tariff but also by the import surcharge. This was originally imposed late in October 1964 at the rate of 15 per cent; it was reduced to 10 per cent after six months and abolished on 30 November 1966. The tariff rates used for the UK in the adjustment of trade balances (see Chapter 3, pp. 25–27) have therefore been raised by a flat 7 percentage points – the average surcharge over all three years. However, the comparisons of UK and EEC tariffs in Chapters 2 and 4 throughout exclude the surcharge.

In the case of the EEC there was the problem that in the years covered by the study – 1965 to 1967 – there were still divergencies in national tariffs on imports from third countries. The EEC's tariff adjustment time-table (as amended in the early 1960's) provided for 60 per cent of the differences between the original

1 Sources: UK: *HM Customs and Excise Tariff Amendment No. 129*, London, December 1966. EEC: CECA/CEE/ Euratom (later Commission des Communautés Européennes), *Tarif Douanier des Communautés Européennes*, July 1963 and subsequent issues.

2 The averaging procedure used was first to combine the rates for any subheadings (or even finer divisions), and then to average the rates for the headings themselves.

national tariffs and the CET being eliminated by 1 July 1963, and for the remaining 40 per cent by 1 July 1968. In general, therefore, during the period of our study the divergencies between national tariffs and the CET should have amounted to 40 per cent of the original differences. However, in practice harmonization of national tariffs had proceeded rather further than the general provision would indicate, for in its June 1967 report the EEC Commission stated that the following proportions of national tariff rates had by that time been equalized to the CET – Germany: 41.4 per cent; Benelux: 23.5 per cent; France: 19.8 per cent; Italy: 32.2 per cent.[3] These rates refer to all commodities (other than foodstuffs) and not just to manufactures, but it appears very unlikely that complete harmonization would be confined to items which are neither food nor manufactures. The Commission did not say anything about the extent of the divergencies which still remained, but inspection of pre-EEC national tariffs suggests that in the great majority of cases the differences were below 5 percentage points. In the light of these considerations it appears that use of the CET to represent national tariffs appears a defensible simplification.

The tariff data used throughout refer to nominal rates. As recent developments in the analysis and interpretation of tariff protection have shown, there are quite strong arguments in favour of calculating effective rates of protection. The data required for this to be done for the 230 commodities do not exist, though given time it might have been possible at least to make some rough adjustments of the nominal rates.[4]

The tariff rates themselves (classified by industry) can be found in Chapter 4, and there is no point in re-stating the information here. However, for reasons explained earlier there has been no occasion to set down the EEC's post-Kennedy Round tariff rates for commodities which the mutual trade rank order would uncontestably assign to class III. For the convenience of readers these rates are therefore shown in Appendix Table B.6.

3 Source: European Economic Community, Commission, *Tenth General Report on the Activities of the Community*, Brussels, 1967, Table 1.

4 Estimates of differences between nominal and effective rates of protection are available both for the UK and the EEC. See T.S. Barker and S.S. Han, 'Effective Rates of Protection for UK Production', *Economic Journal* (forthcoming), and H.G. Grubel and H.G. Johnson, 'Nominal Tariffs, Indirect Taxes and Effective Rates of Protection: The Common Market Countries 1959', *Economic Journal*, December 1967, pp.761–776.

Appendix B:

Supplementary Tables

Table B.1 *Composition of Performance Classes (Mutual Trade Ranking) by SITC Section*

CLASS \ SITC	5	6	7	8
I	533.1 533.2 533.3 541.1 541.3 541.7 554.1 581.9 599.7	611.4 611.9 613.0 621.0 642.9 651.7 653.2 653.3 657.5 663 667.2 676 681.1 681.2 682.2 683.1 683.2 685.1 687.1 694.2	711.4 711.5 712.5 729.1 729.4 729.5 729.6 732.6 732.8 733.3 734	861.4 861.6 862.3 863.0 894.2 894.4
II	512.1 512.2 513.5 571	632.4 641.2 641.9 651.2 651.5 655.4 655.6 655.8 656.1 657.4 662 665.8 671.2 672.5 672.7 673.5 674.1 674.2 674.7 677.0 678.2 678.5 685.2 689 691.1 693.1 695.2 696.0 698	715.2 719.2 719.3 719.4 719.9 723.1 724.9 731 723.3 732.4 732.7	821.0 861.9
III	514.3 521 531.0 553.0 581.3 599.2	629.1 641.1 641.5 642.1 651.3 653.4 653.7 654.0 656.6 657.6 666.4 672.3 673.1 673.2 673.4 674.8 675.0 678.1 678.3 679 682.1 686.2 697.1 697.2	712.1 714.9 718.4 722.1 722.2 725.0 729.9 732.1 732.2 732.5 733.1	841.1 841.4 841.5 861.7 891.2
IV	512.3 512.4 513.2 513.6 514.2 532 554.2 561.2 581.1 581.2	612 641.3 651.6 652.1 653.5 656.9 661.2 665.1 674.3 684.1 684.2 686.1 692.1	712.2 714.2 717.1 717.2 717.3 718.1 718.2 719.1 719.5 719.6 719.7 719.8 723.2 724.1 724.2 729.2	812.4 831.0 841.2 862.4 893.0 895.2 897.1
V	512.5 512.7 512.8 513.3 514.9 551 551.1 561.3 599.5	611.3 631.1 631.2 641.6 651.4 652.2 653.1 653.6 656.2 664.3 664.4 665.2 671.4 671.5 678.4 694.1	711.3 714.1 714.3 715.1 718.3 718.5 726 729.3 732.9 735	841.3 851.0 861.1 861.2 861.3 861.5 864.1 864.2 891.1 891.8 892.9

Notes: (i) Source: Derived from calculations described in Chapter 3.
(ii) For the definition of SITC sections 5-8 see note (ii) to Table 2.2.

Table B.2 *Composition of Performance Classes (Third Country Export Ranking) by SITC Section*

CLASS \ SITC	5	6	7	8
I	521 533.1 533.2 554.1 571 599.7	611.4 611.9 641.2 651.4 651.5 653.2 653.3 656.2 657.6 666.4 667.2 672.3 674.7 679 681.1 681.2 683.2 687.1 696.0	711.3 711.4 711.5 712.1 712.5 714.9 723.1 723.2 724.1 729.1 732.3 732.7 732.8 732.9 733.1 734	861.9 862.3 863.0 892.9 894.2
II	514.2 533.3 554.2 581.1 581.3 581.9	621.0 641.5 641.6 641.9 642.9 651.3 653.4 655.6 655.8 656.9 657.4 657.5 663 664.4 665.8 676 678.5 683.1 686.2 693.1 695.2 697.2 698	717.1 717.3 718.4 719.3 719.7 719.9 722.2 724.9 729.5 729.6 731 732.4 732.6	821.0 861.6 894.4 893.0
III	513.6 514.3 531.0 541.3 541.7 553.0 581.2	612 629.1 651.7 652.1 652.2 654 655.4 661.2 673.4 674.2 675.0 678.1 682.2 685.1 691.1 694.2	712.2 714.2 715.1 717.2 718.1 718.3 718.5 719.1 719.2 719.5 719.6 719.8 722.1 725.0 729.4 729.9 732.1 732.2	841.1 841.5 861.7 891.1 891.2
IV	512.1 512.2 512.4 512.5 514.9 551 599.2 599.5	613.0 642.1 651.2 651.6 653.5 653.7 656.6 665.1 671.5 672.5 674.1 674.3 674.8 677.0 678.2 682.1 684.1 684.2 685.2 689 692.1 697.1	714.3 715.2 718.2 719.4 729.2 729.3 732.5 733.3 735	812.4 841.2 851.0 861.1 861.4 862.4 864.2
V	512.3 512.7 512.8 513.2 513.3 513.5 532 541.1 561.1 561.2 561.3	611.3 631.1 631.2 632.4 641.1 641.3 653.1 653.6 656.1 662 664.3 665.2 671.2 671.4 672.7 673.1 673.2 673.5 678.3 678.4 686.1 694.1	714.1 724.2 726	831.0 841.3 841.4 861.2 861.3 861.5 864.1 891.8 895.2 897.1

Notes: (i) Source: Derived from calculations described in Chapter 3.
(ii) For the definition of SITC selections 5-8 see note (ii) to Table 2.2.

Table B.3 *Conflicting Evidence Products with Rank Order Difference of 91 Steps or Less*

Class by Mutual Trade Ranking	SITC	Description	Class by Third Country Export Ranking
I	682.2	Copper, worked	III
	685.1	Lead, unwrought	III
	694.2	Nuts & bolts	III
	729.4	Electrical equipment for vehicles	III
II	512.1	Hydrocarbons & derivatives	IV
	512.2	Alcohols	IV
	672.5	Blooms & billets	IV
	677.0	Wire	IV
	715.2	Other metal working machinery	IV
III	521	Mineral tar	I
	657.6	Other carpets	I
	672.3	Ingots	I
	673.1	Wire rod	V
	673.2	Bars & rods	V
	679	Castings	I
	714.9	Other office machinery	I
	841.4	Knitwear	V
IV	514.2	Other metallic salts (I)	II
	554.2	Detergents	II
	717.1	Textile machinery	II
	717.3	Sewing machines	II
	719.7	Ball bearings	II
V	715.1	Machine tools	III

Note: For an explanation of this table see Chapter 3, pp. 33–34.

Table B.4 *Class Distribution of Products by Number (Third Country Export Ranking)*

per cent

Industry	No. of Products	Class I	II	III	IV	V	Total
Chemicals	38	15.8	15.8	18.4	21.1	28.9	100.0
Manufactures of leather, rubber, wood, paper, etc.	32	15.6	28.1	9.4	12.5	34.4	100.0
Textiles and clothing	30	20.0	23.3	23.3	20.0	13.3	100.0
Iron and steel	24	12.6	8.4	16.8	29.2	33.3	100.0
Non-ferrous metals and metal manufactures	24	20.9	25.0	16.7	29.2	8.3	100.0
Mechanical engineering	29	20.6	24.1	37.9	13.8	3.4	100.0
Electrical engineering	19	21.1	21.1	31.6	15.8	10.5	100.0
Transport equipment	14	43.9	26.4	14.3	21.4	0	100.0
Miscellaneous manufactures	20	25.0	10.0	10.0	20.0	35.0	100.0
Total	230	20.0	20.0	20.0	20.0	20.0	100.0

Source: Appendix Table B.2 and Tables 4.A.1, 4.B.1, etc.

Table B.5 *Class Distribution of Products by Value (Third Country Export Ranking)*

per cent

Industry	As % of total intra-trade in manufactures	Class I	II	III	IV	V	Total
Chemicals	11.6	6.8	19.2	29.7	22.2	22.1	100.0
Manufactures of leather, rubber, wood, paper, etc.	10.3	22.7	29.6	9.9	14.6	23.2	100.0
Textiles and clothing	13.2	17.9	6.5	28.3	30.1	17.1	100.0
Iron and steel	10.1	4.5	2.7	15.1	40.8	36.9	100.0
Non-ferrous metals and metal manufactures	8.9	13.6	26.1	18.4	39.2	2.8	100.0
Mechanical engineering	19.3	20.7	28.1	39.0	10.7	1.5	100.0
Electrical engineering	9.5	7.7	32.0	39.1	13.9	7.4	100.0
Transport equipment	12.5	41.9	1.9	48.0	8.2	0	100.0
Miscellaneous manufactures	4.6	34.6	5.8	18.0	24.3	17.3	100.0
Total	100.0	18.2	17.4	31.2	20.5	12.6	100.0

Source: Appendix Table B.2 and Tables 4.A.3, 4.B.3, etc.

Table B.6 *Post-Kennedy Round CET Rates of Uncontested Class III Products*

SITC	Description	Post-Kennedy CET (per cent)
514.3	Other metallic salts (II)	9.6
531.0	Synthetic dyes	12.8
553.0	Cosmetics	12.0
581.3	Cellulose and derivatives	11.2
599.2	Insecticides, etc.	7.7
629.1	Rubber tyres	8.3
641.5	Machine-made paper	12.0
642.1	Paper bags, etc.	15.0
651.3	Grey cotton yarn	8.0
653.4	Jute fabrics	20.0
653.7	Knitted fabrics	13.3
654	Small wares	10.5
656.6	Blankets	13.0
841.1	Clothing excluding knitwear	15.2
841.5	Headgear	8.4
673.4	Angles etc. 80 mm or more	6.7
674.8	Thin coated plates	6.7
675.0	Hoop & strip	6.7
678.1	Cast iron tubes	9.0
682.1	Copper, unwrought	0
686.2	Zinc, worked	8.7
697.1	Stoves & cookers	7.0
697.2	Domestic utensils	7.5
718.4	Construction machinery	6.3
722.1	Electrical power machinery	6.0
722.2	Switchgear, etc.	6.8
725.0	Domestic electrical equipment	6.7
729.9	Other electrical machinery	6.3
891.2	Records & tapes	6.7
732.1	Passenger motor cars	14.5
732.2	Buses	14.5
732.5	Road tractors	18.0
861.7	Medical instruments	7.3

Notes: (i) Source: See Appendix A.
(ii) Post-Kennedy CET rates of conflicting evidence class III items are shown in Tables 4.A.4, 4.B.4, etc.